CAMBRIDGE UNIVERSITY: NEW ENGINEERING WORKSHOPS.

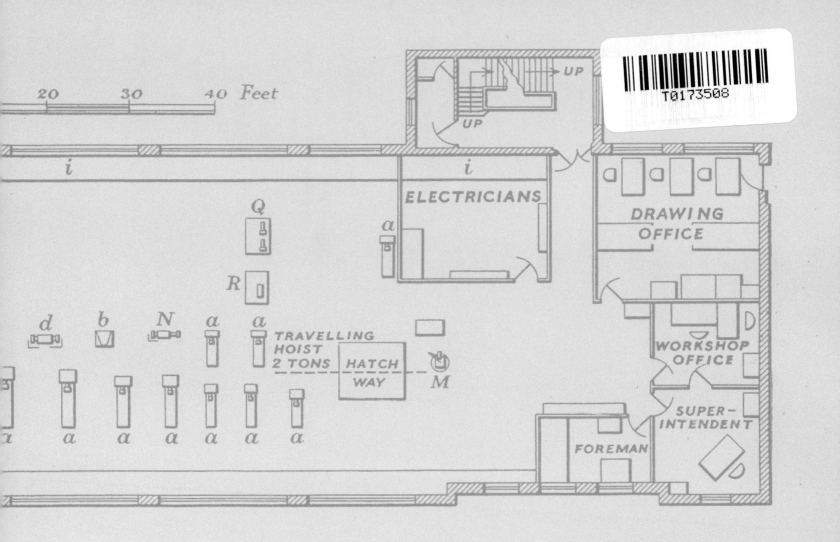

a	Lathe.	m	Gear shaper.	y	Bending roller.	K	Straight-line cutter.
b	Drilling machine.	n	Surface grinder.	z	Setting-out table.	L	Brazing hearth.
c	Shaper.	o	Plain grinder.	A	Capstan lathe.	M	Engraving machine.
d	Grinder.	p	Guillotine.	B	Radial drilling machine.	N	Buff and grinder.
e	Planer.	q	Nibbler.	C	Horizontal boring machine.	P	Demonstration welding bench.
f	Forge.	r	Mortise machine.	D	Vertical boring machine.	Q	Watchmaker's lathes.
g	Cupboard.	s	Trimmer.	E	Power saw.	R	Coil-winding machine.
h	Linisher.	t	Shears.	F	Spot welder.	S	Slotting machine.
i	Bench.	u	Band saw.	G	Argon-arc welder.	T	Planing machine.
j	Multi-drilling machine.	v	Circular saw.	H	Butt welder.	U	Belt sander.
k	Milling machine.	w	Fly press.	J	Profile cutter.	V	Tool grinder.
l	Universal milling machine.	x	Screwing machine.				

CAMBRIDGE ENGINEERING
THE FIRST 150 YEARS

CAMBRIDGE ENGINEERING
THE FIRST 150 YEARS

Haroon Ahmed

Third Millennium
Publishing

UNIVERSITY OF
CAMBRIDGE

Department of Engineering

First published in Great Britain in 2017 by
Third Millennium Publishing, an imprint of Profile Books Ltd.

3 Holford Yard, Bevin Way
London WC1X 9HD
United Kingdom
www.tmbooks.com

A CIP catalogue record for this book is available from The British Library.

ISBN 978 1 908990 68 6

Project Manager: Susan Millership
Design: Matthew Wilson
Production: Simon Shelmerdine

Reprographics by Studio Fasoli, Italy
Printed and bound in China by 1010 Printing International Ltd
paper from sustainable forestry

Previous pages: PhD student Charlotte Coles setting up an
experiment in the supersonic wind tunnel in the Baker Building.

Above: Edmund Kay, Alistair Gregory and John Hazlewood
in the Acoustic Laboratory's anechoic chamber.

Contents

Foreword

Today, we take engineering for granted. It is everywhere. Almost everything you see and touch is the product of engineering – this book you are holding now was produced using an array of forestry machinery, a modern paper mill, high-resolution digital cameras, the latest computers, communications networks, precision-engineered printing processes, an optimised transport network. Engineering's ubiquity, curiously, tends to make it invisible.

This book brings engineering into sharp relief and celebrates Cambridge's role in its development. One hundred and fifty years ago, engineering struggled to find its feet in the University amid academic opposition to this new discipline. Now, Engineering is the largest department – a powerhouse of research, teaching and beneficial global impact.

The story is one of astounding change. In 1878, the Department was entirely contained within a 15-by-6-metre hut; it is now poised to create 100,000 square metres of the world's finest engineering facilities at West Cambridge. For decades from its start in the nineteenth century, engineering was the preserve of men from a narrow range of backgrounds; now the Department is home to staff and students from around the world and women are leading the way in many fields.

There is much more still to do. The Department must gather and nurture the finest engineers in the world to address the world's many challenges, including global climate change, economic growth and the ageing population. We

Engineering students working on an aeroplane in the early 1920s.

are perfectly positioned to make a tremendous contribution with the world's most brilliant students, superb faculty and excellent support staff working together in a single integrated department. The planned new facilities will enable us to take Engineering to the next level in partnership with other leading institutions, government and industry.

The history of engineering told in this book shows that Cambridge Engineering has come a long way in nearly 150 years. Its reach and impact continues to accelerate. Even my wildest predictions for the next 150 years are likely to fall short of what this unique international community of engineers will achieve.

Ann Dowling
November 2016
Dame Ann Dowling is President of the Royal Academy of Engineering and was Head of the Department of Engineering from 2009 to 2014

Above: Ann Dowling with James Dyson (right) and his colleagues, Charles Collis (left) and Frederic Nicolas (second from right), reviewing collaborative research in the Acoustics Laboratory.

Below: A student demonstrates with LEGO at an outreach event in the Department's Design Project Office.

Acknowledgements

Engineering teaching began in Cambridge early in the nineteenth century, well before the foundation of the Engineering Department in 1875. The fascinating history of the subject, from its origins to the present day, is recorded in *Cambridge Engineering: The First 150 Years*. I am most grateful to David Cardwell, the Head of the Engineering Department, for giving me the opportunity to write this book, and for his support throughout the project. I would also like to acknowledge my close cooperation with Keith Glover whose advice and guidance were invaluable. Jacques Heyman's encouraging remarks after 'speed reading' the whole of the original manuscript enabled me to complete the work in just two years. Former senior members of the Department, Nick Cumpsty, Chris Calladine, Andrew Schofield and Donald Green must also be acknowledged for their contributions to this work.

Invaluable reference sources for the early days of engineering at the University included Elisabeth Leedham-Green's *A Concise History of the University of Cambridge*, and *A History of the University of Cambridge* by C N L Brooke. The founding of the Engineering Department by James Stuart and the contributions of his successors, James Alfred Ewing, Bertram Hopkinson and Charles Inglis are recorded in detail by T J N Hilken in *Engineering at Cambridge University 1783–1965*, and in articles in journals of the Cambridge University Engineering Society (CUES). Ewing, unlike Stuart, was adroit in University politics and he succeeded in establishing the Mechanical Sciences Tripos. His successor, Hopkinson, not only raised the standard of the Tripos but also instituted the ethos of research into the Department. For the period when Inglis served as Head of Department,

Opposite: The Cambridge University High Resolution Electron Microscope was a collaboration between the Department of Engineering and the Cavendish Laboratory. Initial results from this 600 kV transmission electron microscope, showing atomic resolution, were published in *Nature* in 1979.

Below: Technicians from the Trumpington Street Site in the Baker Building foyer in 2016, with the Suspension Bridge mural by Tony Bartl (1912–1998) as the backdrop.

HRH The Duke of Edinburgh, on the right, looking at a model of the Baker Building at its opening in 1952.

POEM WRITTEN BY AN ALUMNUS

The poets pray their verses will survive
So long as man can breathe or eyes can see
And scientists bestow their names upon
Ideas in hope of immortality
But engineers confront what lies before
In challenges pursuing nothing more
Than elegant solutions built on proof
Whose working not naming enshrines the truth

Anon

of Department, when teaching reigned supreme and research was largely neglected, much useful information is available in the journals of the CUES and in Hilken's comprehensive treatise.

The Annual Reports to the University instituted by John Baker were the most significant sources of information for his headship as were those of William Hawthorne, Austyn Mair, Heyman, Alec Broers and David Newland. Heyman and Broers presented the case for more space for the Engineering Department with increasing passion and frustration in their reports. The Department's submissions to the Research Assessment Exercise and the Research Excellence Framework were the main sources for information on research activity when Newland, Glover and Ann Dowling were Departmental Heads. The Department Librarian, Niamh Tumelty, and the other library staff are thanked for their helpful guidance through the library archives.

Broers, Newland and Glover provided helpful comments on the chapters covering their headships and Robert Mair read the section on his father's term in office. Chapter 8 was written by Claire Barlow and several members of staff helped me to write the penultimate chapter covering some of the many highlights of research in the Department. David Cardwell wrote the first draft of the final chapter 'Going West'.

Particular thanks are due to Peter Long, Alistair Ross and John Harvey for access to their personal and archival picture collections. Alan Davidson of STILLS Photography is thanked for providing many of the excellent photographs in this book. The project editor, Susan Millership, and the designer, Matt Wilson, are acknowledged for their help with the book. Finally I am grateful to my wife, Anne, for reading the first drafts and for making numerous constructive suggestions which greatly improved the manuscript.

Haroon Ahmed
December 2016

CHAPTER ONE

The Origins of Cambridge Engineering

The Industrial Revolution and the advent of powered machinery in the nineteenth century heralded the demise of cottage industry. Across the nation, the power of water and steam was harnessed for manufacturing as the production of goods was transferred to purpose-built factories and new national, as well as international, markets became available for mass-produced goods. It was manifestly evident that technical and scientific training and expertise would enhance the performance of the labour force, improve manufacturing processes and increase profitability.

In the midst of this era, the gratifying news was received in Cambridge that the Reverend Richard Jackson (1700–1782), a former Fellow of Trinity College, had left monies in his will to endow a professorship in the University, with one of the stipulations being that the incumbent should strive for 'the promotion of real and useful knowledge' and use experiments to illustrate his teaching. The Jacksonian Professorship of Natural Experimental Philosophy was thus founded, and holders of this Chair have had a profound influence on Cambridge science and engineering. Three Jacksonian Professors, Isaac Milner, William Farish and Robert Willis, along with the Astronomer Royal, George Biddell Airy, established the teaching of 'real and useful knowledge', or in today's terms 'engineering', at the University of Cambridge.

But much earlier in the history of the University of Cambridge, the arrival of Isaac Newton (1642–1727) in 1661 at Trinity College had given the University an edge over its great rival Oxford in mathematics and science. (Roy Jenkins, Chancellor of Oxford University and graduate of Balliol College, Oxford, while speaking in the Senate House of the University of Cambridge bemoaned the favours bestowed upon Cambridge by Newton.) Newton's scientific discoveries astounded the world and Newtonian mathematics had a significant influence on Cambridge teaching.

Robert Willis (1800–1875), Jacksonian Professor from 1837 to 1874, described by his biographer as 'the archetypal nineteenth-century polymath', was an eminent mathematician who taught engineering before the foundation of the Department.

NOBLEMEN

Dress Gown.   Undress Gown.

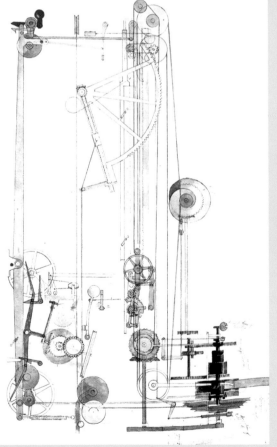

Above: Roubiliac's statue of Sir Isaac Newton (1642–1727) stands in the ante-chapel of Trinity College, Cambridge, where Newton was a Fellow.

Above left: Illustrations of noblemen's attire from *The Costumes of the Members of the University of Cambridge*.

Left and right: Extracts from Robert Willis' remarkable scrapbooks.

Below: A letter opener presented to James Stuart, the founder and first head of the Engineering Department from 1875 to 1889, at his leaving party in 1890.

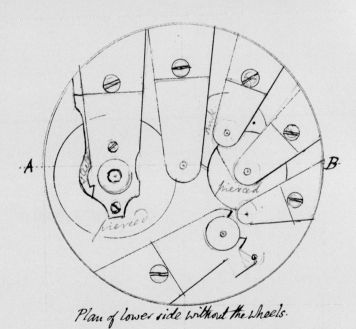

Plan of lower side without the wheels.

Presented to PROFESSOR STUART BY THE Cambridge University Engineering Department JULY 1890

Newton was admitted to Trinity College as a sizar, a term used at Cambridge for a student who received financial assistance. This status required him to perform menial tasks around the College for the comfort of fellows and 'noblemen undergraduates' in return for his education, board and lodging. He survived this hardship, and also the Black Death in 1666 by fleeing temporarily to his family home, Woolsthorpe Manor in Lincolnshire; the plague took around a third of Cambridge's citizens to their graves. At Trinity, Newton very quickly established a remarkable reputation as a mathematician and scholar, and when Isaac Barrow, Lucasian Professor of Mathematics at the University of Cambridge, Newton's teacher and Fellow of the Royal Society, resigned his Chair he recommended that Newton, then just 27 years of age, should fill the vacancy.

The publication of Newton's great work, *Principia*, in 1687 and the acclaim that followed ensured that mathematics would become an essential course of study in the University of Cambridge. Thereafter, for some 200 years, many scholars in the University held the firm belief that mathematics was the 'honour and glory' of Cambridge teaching and any introduction of useful knowledge had to be based on this rigorous discipline. Newtonian mathematics dominated the teaching of undergraduates, even though the majority were preparing to take the Ordinary Degree and going on to take Holy Orders. Subjects such as the natural sciences or classics could only be studied, by the more able men, after the prescribed course in mathematics, sometimes referred to as the 'Tripos', had been successfully completed. The main textbooks were Euclid's *Geometry*, Newton's *Principia*, *On Human Understanding* by Locke and *Moral Philosophy* by Paley. Jacksonian Professors were thus constrained to design their lecture courses on a mathematical foundation even when they were teaching some of the practical elements of engineering.

MATHEMATICIANS AND ENGINEERS

Also admitted as a sizar, Isaac Milner (1750–1820) was to go on to become the Founding Jacksonian Professor in 1783. Milner had graduated as Senior Wrangler (the top undergraduate in mathematics), and won the Smith's First Prize, which was awarded to the Cambridge student who had made the greatest progress in mathematics or natural philosophy. These achievements were rewarded with a college tutorship and an academic career in the University of Cambridge, where Milner's mathematical research gained him Fellowship of the Royal Society in 1780. Three years later he was the outstanding candidate for the Jacksonian Chair.

An engineer at heart, Milner introduced into his mathematical lectures the theory of the steam engine, descriptions of air pumps

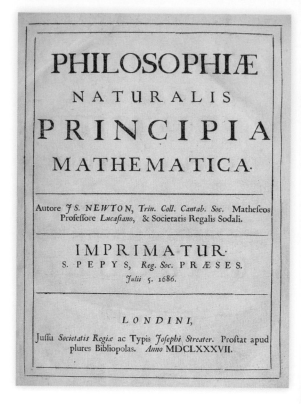

Portrait from Queens' College of Isaac Milner, first Jacksonian Professor from 1783 to 1792, who started teaching the rudiments of engineering within mathematics lectures.

Opposite: The publication of Newton's *Philosophiæ Naturalis Principia Mathematica* had a profound influence on the teaching of undergraduates at Cambridge University for more than a century.

William Farish, chemist and Jacksonian Professor from 1813 to 1837, taught construction and use of machines.

and elementary mechanisms. His scientific eminence led to his election to President (Head of the College) of Queens' College, and thereby a home where he was able to build a workshop and install in it workbenches and machinery such as lathes and drills 'to fashion useful mechanical devices and to conduct diverse electrical and mechanical experiments' for the next 32 years. His reputation as a practical 'engineer' became widespread, and he was asked by the government of the day to advise on the construction of a new bridge across the River Thames – probably the first recorded case of a Cambridge professor acting as a consultant! To Milner goes the credit of being the first professor to introduce 'real and useful knowledge' into his mathematical teaching, thus honouring the terms of Jackson's legacy. Beyond this activity, the publication of his seven-volume *Ecclesiastical History of the Church of England* and his influence on William Wilberforce for the abolition of slavery made him renowned across the nation.

William Farish (1759–1837), Jacksonian Professor (1813–1837), was 'the first man to teach the construction and use of machines as a subject in its own right'. Admitted to Magdalene College as a sizar, he also graduated as Senior Wrangler and was duly ordained and appointed to a College tutorship. Farish delivered his lectures entitled the 'Arts and Manufactures of this Kingdom' in the Jacksonian Lecture Rooms while holding the Professorship of Chemistry (1794–1813). Later, as Jacksonian Professor, he produced a plan in which he described the extraction, processing and use of metals in making finished products, as well as considering non-metallic substances used in the making of glassware, window panes and optical lenses. The plan included a comprehensive section on the construction of machines designed for the extraction of power from wind, water and steam, and he also explained how such power could be harnessed for the use of man, going on to show that the principles he was teaching were essential to understanding the operation of steam engines, boilers and air pumps. In his comprehensive lectures, often supported with working models, he described gauges, valves, pistons, gear wheels, lathes and lifting mechanisms.

Farish was so celebrated a lecturer that he attracted 'very crowded audiences', and it was said of Farish's teaching quality that a student attending his lectures 'will carry with him into life so much ingenious knowledge, if he has given his attention to the course, that he will everywhere meet with consideration and respect, while he can render service or furnish instruction' – an apt characterisation of an engineer serving the community, both then and now. In the nineteenth century, as steam engines became commonplace sources of motive power and technology advanced rapidly, Farish's lectures became outdated, but in his audience was Robert Willis, who would take on the mantle of Jacksonian Professor in 1837 and continue to lecture on 'real and useful knowledge'.

Willis (1800–1875), probably the greatest teacher of engineering within the Cambridge mathematics syllabus and a 'man of astonishing versatility', was elected Jacksonian Professor following the death of Farish in 1837. On graduating

AIRY HEIGHTS: FROM CATHEDRAL VAULTS TO THE CANOPY OF STARS

George Biddell Airy (1801–1892), admitted to Trinity College as a sizar, established a reputation as an outstanding scholar and teacher while still an undergraduate, and immediately after his graduation as Senior Wrangler and first Smith's Prizeman the College appointed him to an assistant mathematical tutorship and a fellowship. At the age of 25 he was elected to the Lucasian Professorship, with a stipend of £99 per annum, but vacated the chair in favour of the better paid (£300 per annum, later increased to £500 per annum) Plumian Professorship of Astronomy and Experimental Philosophy (1828–1836).

Airy lectured in a small room under the University Library on 'engineering topics' in the guise of Experimental Philosophy. His syllabus comprised statics and dynamics, hydrostatics, pneumatics, optics, gear wheels, screws and pulleys, but his main interest was in structures; his popular lectures on the theory of roofs were illustrated with examples drawn from the domes of the cathedrals of Rome and St Paul's in London and from the fan vaulting in the roof of King's College Chapel.

Airy was appointed Astronomer Royal at the Greenwich Observatory in 1835. He re-engineered many of the telescopes and other instruments at the Observatory and supervised their construction. He also introduced the use of electricity for signalling the time from Greenwich across the country. The first electric signal was sent from Greenwich to Lewisham in 1852. The government sought his advice on a wide range of issues, such as the standardisation of the gauge of railway line and signalling, but today his judgement is considered to have been occasionally flawed; in particular, he became a prominent detractor of the work of Charles Babbage, advising the government against funding the completion of the inventor's remarkable Difference Engine.

George Biddell Airy, Astronomer Royal from 1835 to 1881, is depicted as a time ball, a device that provided visual time signals for shipping. Airy was the driving force behind extending the country's network of time balls and controlling them with electrical signals from Greenwich. Earlier in his career he included engineering topics in his lectures at Cambridge.

from Caius College he was elected to a fellowship and ordained. However, he had no intention of becoming a clergyman and immediately took up his preoccupations with mechanisms, devising and constructing models, some with intricately intermeshed gear wheels, and making many other mechanical contrivances. His books *Principles of Mechanism* and *A System of Apparatus for the Use of Lecturers and Experimenters in Mechanical Philosophy*, as well as his course of lectures entitled 'A Course of Experimental Lectures on the Principles of Mechanism', demonstrate his consuming passion for mechanical engineering.

Willis, a lecturer of exceptional ability, delivered stimulating lectures 'using neither manuscript nor notes', and gave fascinating demonstrations of complicated mechanisms and models that he made himself in his workshop. A former student described his lectures as follows: 'An uninterrupted stream of lucid exposition flowed from his lips, carrying his hearers without weariness through the most intricate details and making them grasp the most complex history of construction.' Undergraduates flocked to his lectures in large numbers, deserting their college tutors, who necessarily confined their teaching closely to the relatively dull topics of the prescribed mathematics syllabus.

Willis, described by his biographer Alexandrina Buchanan as 'the archetypal 19th-century polymath', did not confine himself to teaching in Cambridge, but also taught at the Royal School of Mines in London (now subsumed into Imperial College) and maintained links with other scientists and engineers, including the great civil engineer Thomas Telford. He was also a pioneer in architectural history; in 1862 he was awarded the Gold Medal of the Royal Institute of British Architects and he is acknowledged today as perhaps the greatest architectural historian.

The Royal Commission of 1850 (see box overleaf entitled 'The Reformers'), mindful of Willis' engineering expertise and the quality of his lectures, proposed that the University of Cambridge should establish a Professorship of Engineering as well as a Tripos in the subject. Willis was clearly the best candidate for this position, though the proposal was not pursued in the University due to lack of funds. Even if funds had been available, Willis may not have been tempted to give up the Jacksonian Chair in favour of the task of setting up a new department, with all its inevitable administrative duties.

The University moved gradually towards accepting engineering as a necessary discipline, and the topic of 'mechanism and applied science' was included among the 'special examinations' introduced in 1865, with Willis delivering many of the lectures himself with James Clerk Maxwell sharing the load on his election as Cavendish Professor of Physics in 1871. Willis died in 1875, coincidentally the year in which the University at long last established a Professorship in Mechanism and Applied Mechanics, with the financial assistance of Trinity College in the payment of a stipend of £300 per annum to the Professor, whose duties required him to teach the principles of mechanism, the theory of structures, the theory of machines and the steam engine and other prime movers. The lectures had to be approved by the Board of Mathematical Studies, and it was not clear at this stage whether creating the Professorship implied the establishment of a School of Engineering, but opponents to an 'engineering' professorship, who lost the vote in the Senate House by 74 votes to 36, certainly hoped to find another opportunity to put an end to the teaching of the subject. A prominent opponent, Robert Phelps, wrote: 'However, I am happy in having publicly protested.'

Engineering, and indeed many of today's mainstream subjects, may never have been taught in Cambridge and Oxford if groups of concerned individuals had not come to the conclusion in the early years of the nineteenth century that 'Oxford and Cambridge had failed in their duty to promote the advancement of learning and they were incapable of reforming themselves'. Cambridge college teaching was dominated by the needs of the Church of England, with most Cambridge graduates going on to become Anglican clergymen. The standard of entry was abysmally low except for sizars, who had to demonstrate exceptional talent to gain fee-free entry and the right to earn their keep by serving their social, although not remotely their intellectual, 'betters'.

READING AND NON-READING MEN (VARMINTS)

Unofficially the undergraduate body was divided into the serious-minded and studious, or 'reading men', and the 'non-reading men', also known as 'varmints', who were noblemen or wealthy laymen destined for a life of ease and prosperity. With just a smattering of education acquired and very little study, their time during the three years was spent primarily in excessive drinking, gambling, hunting, fishing and shooting in the pleasant surroundings of Cambridge.

'Varmints' taking part in the 'old varsity trick' of smuggling a lady into college.

THE REFORMERS

In the first stage of reformation, predating engineering, a Previous Examination known as 'Little-go' was made compulsory in 1822, thus raising the standard of those admitted to the University by reducing the number of 'non-reading men'. In a second stage, the emphasis on mathematics was diminished when the Mathematical Tripos was divided into two parts in 1848, with Part I set at a comparatively simple level to be attempted by all entrants, enabling many to move on to subjects of study other than mathematics, whereas Part II was maintained at a demanding level appropriate for those with high academic aspirations.

These developments, although welcome, did not go far enough, and there was concern that the University and the 17 colleges were continuing to serve as seminaries for the Church of England rather than serving the needs of the nation. Fortunately, in the middle of the nineteenth century, two men who possessed great power and authority, Prince Albert, consort of Queen Victoria and also Chancellor of the University, and Lord John Russell, the Prime Minister, came together to overcome the intransigence of the colleges as well as the machinations of the University's diehard conservatives.

Russell approached the task by way of a frontal attack, proposing a Royal Commission of Inquiry, while the Chancellor adopted a more conciliatory stance, by beginning negotiations with his academic friends. Neither man had studied at Cambridge, but both were aware of the complexities of Cambridge, with its independent colleges, each with its own charter and statutes.

The Chancellor was able to encourage some minor reforms but Russell was not satisfied with this tinkering and his hand was greatly strengthened in July 1848 when he received a petition signed by 133 Cambridge graduates, including such notables as Charles Darwin, William Thackeray and Charles Babbage, urging vigorous action on reform. Russell exercised his constitutional authority in 1850 and appointed a Royal Commission of Inquiry. The news was met with indignation in Cambridge, where it was considered an affront and an indignity that such an ancient university should be subjected to an external review.

The Commissioners proposed a number of reforms, asking for new Triposes to be instituted in law, engineering and theology, recommending that Professors be appointed in Practical Engineering and Descriptive Geometry, and causing consternation in the colleges by suggesting that professorial salaries should be paid by a levy upon the colleges. These reforms, although welcome, were still not radical enough to satisfy Russell, the main stumbling block being the University's 'veneration' of the study of mathematics. Among the colleges it was feared that new subjects would make it impossible for

Lord John Russell (1792–1878), leading Liberal politician and principal architect of the Reform Act of 1832, also led the movement for the reform of Cambridge University. He appointed a Royal Commission in 1850 that promulgated new Statutes in the Cambridge University Act of 1856.

Portrait of Prince Albert by Frederick Richard Say, commissioned to mark his election to Chancellor of the University of Cambridge in 1847. He supported changes that helped steer the University from being primarily a 'seminary for the Church of England' to a national university.

college tutors to teach and examine undergraduates in the customary manner, and that changes in the syllabus could require them to acquire new knowledge.

Pressure from the reformers was met with studied dilatoriness in Cambridge, and the University remained largely unchanged, but in 1854 Russell returned to the attack, demanding a more effective and democratic form of government within the University. Specifically he thought that fellowships should be given only to men prepared to teach undergraduates and do research, that students should have the freedom to attend the teaching provided by the University professors and finally that restrictions which designated special favours upon a particular locality or institution linked to a college should be removed. Fly sheets were circulated by both opponents and supporters of the reforms and there were vicious 'paper wars' but no significant progress. The argument was put forward that subjects such as engineering and chemistry would be best taught at universities that existed in cities where industry was well established and that the intellectual purity of Cambridge teaching should not be tarnished.

Russell set up a Statutory Commission to oversee the reform of the University of Cambridge, and 'An Act to make further Provision for the good Government and Extension of the University of Cambridge' was given royal assent. Over the years that followed, many reforms were carried out: the Caput Senatus, a small executive committee with dictatorial powers (each member held an individual right of veto) and a source of constant hindrance to reform, was abolished and replaced with a democratic Senate; the range of subjects was broadened to produce graduates capable of making careers in industry and commerce; University lectures became an integral part of the teaching of undergraduates; and the election process of college fellows was reformed and the constraints of celibacy removed, reducing the number of clerics produced by the University. A bill was introduced in Parliament to abolish religious tests for all except Heads of Colleges and candidates for divinity degrees, which opened the University to Roman Catholics, non-believers and dissenters. Over the last two decades of the nineteenth century the transformation of the University of Cambridge from a seminary for the Church of England to a national institution of education, scholarship and research was completed.

James Stuart (1843–1913)

Professor of Mechanism and Applied Mechanics (1875–1889);
Head of Engineering (1875–1889)

FOUNDATION OF THE ENGINEERING DEPARTMENT

The bells of the University Church, St Mary the Great, rang out on 17th November 1875 to celebrate the election of James Stuart to the Professorship of Mechanism and Applied Mechanics, marking his triumph over the redoubtable 41–year-old Edward Routh, acknowledged teacher, notable mathematician, Fellow of Peterhouse and the Royal Society, who was defeated by Stuart's 111 votes to his 86. Escorted to the Senate House by a group of his friends and supporters, Stuart knelt before the Vice-Chancellor, took the oath of office in Latin ending with *Do ita fidem* (meaning 'So I swear'), and Engineering in the University of Cambridge was thus founded. At the time of his election, at the age of 32, Stuart was a Fellow and Assistant Tutor at Trinity College and, as Professor of Mechanism and Applied Mechanics, he established the teaching of engineering in Cambridge and nearly succeeded in raising the subject to the level of an honours degree.

On his appointment Stuart was allocated partial use of the Jacksonian Lecture Rooms and two small adjacent rooms on the New Museums site, which became the Engineering Department. He was pleased to inherit the models designed and built by Willis for the teaching of mechanisms, but they were in a state of considerable disrepair. Stuart's first task was to restore them for use in demonstrations at his inaugural lecture. Next he converted one of his small rooms into a workshop, which he equipped with machines and hand tools, enabling him to instruct a maximum of seven students. In these premises he gave lectures on mechanism and applied science, while in the adjacent Cavendish Laboratory his students attended lectures on light, heat, magnetism and electricity given by James Clerk Maxwell, who had been appointed to the Cavendish Chair four years earlier. The first examination in engineering was held in the Michaelmas Term, 1876.

Stuart, determined to increase the number of students studying engineering, persuaded the University to build a wooden hut, 50 feet long by 20 feet wide, to give him a workshop large enough to accommodate up to 25 students. This hut was the first recorded purpose-built Engineering Departmental building, and the opening of the workshop in Cambridge attracted a great deal of attention, reported in detail in national and foreign newspapers and referred to by the Vice-Chancellor in his annual address to the University. Among the notable visitors to the workshops were the scientist

Above: James Stuart, Founder of the Engineering Department, educationist, social reformer and politician.

Below: William Gladstone (1809–1898), Liberal statesman and author, was an early visitor to the Department's workshops.

Charles Darwin and the politician William Gladstone. Fortunately for Stuart this was a period of expansion in the University, which was responding belatedly to the findings of the Royal Commission, and in 1881 a new wing was added to the workshop that made it possible to teach engineering drawing, and small engines were installed for teaching purposes. A year later, Stuart was able to expand his premises by adding a carpenter's shop and a smithy in two cottages allocated to him by the University; the smoke from the smithy damaged plants in the adjacent Botany School, causing much annoyance to the Professor of Botany!

Stuart recruited workshop assistants, an instrument maker, a fitter and a cabinet maker, and the University agreed that a University demonstrator should be appointed to work in the nascent Engineering Department; John Ambrose Fleming, a graduate of St John's College, was the first demonstrator. The syllabus was enlarged and a class in surveying and levelling was initiated. (The teaching of surveying as part of the Engineering Tripos is maintained to this day: groups of weary students can be seen returning to the Department after a long day of surveying Coe Fen, probably the most surveyed piece of land in the UK.)

James Samuel Lyon succeeded Fleming as demonstrator in 1882, adding to the range and diversity of the subjects taught by building a foundry. In 1884 Stuart asked for an additional post of Superintendent of the Workshops with the status of a reader, and Lyon was duly promoted to this position. Three of Stuart's early students were appointed as demonstrators, with salaries paid from fees and profits earned by operating the workshop as a business.

Under Stuart's guidance the number of pupils reading engineering increased rapidly, reaching 70 by 1884, and new subjects such as the design of bridges, the foundations of viaducts and sewage engineering were added to the syllabus. Stuart was by this point a considerable figure within the University and at the zenith of his career; he was a member of three of the most influential bodies governing the University: the General Board, the Council of the Senate and the Board of Physics and Chemistry. His achievements were extraordinary and had no previous parallel in the history of the University – single-handedly he had created a whole new department.

But there was a curious anomaly in the founding of the workshop: the University had provided the funds for the building but Stuart purchased and installed the workshop equipment such as metal turning and cutting machines, hand tools and workbenches at his own expense. It was an unusual arrangement, and one that was to haunt him and eventually bring to an end his career in the University.

In his report to the University in 1883, Stuart asked the University to take over full ownership of his workshop, for a sum of £2,500. The University agreed to purchase the equipment and to run the workshop formally as a University department, with independent valuation experts estimating that a sum of about

Charles Darwin (1809–1882), another eminent visitor to the department. This cartoon from *The London Sketch* of 1874 shows Darwin holding up a mirror to an ape to show how alike they are.

EDUCATION FOR ALL

Stuart became passionately interested in providing resources for the 'vast masses who desire education', and as Secretary of the Local Lectures Syndicate in Cambridge he laid the foundation for the Institute of Continuing Education at Madingley Hall. Using his efforts at Cambridge as an example, he tried to persuade all universities to provide extramural education to a wide cross-section of society. His biographer, Colin Matthew, wrote: 'He should not, as has sometimes been the case, be seen as the sole originator of University extension, but he was certainly its most prominent early activist.'

£4,000 should be paid to Stuart for his equipment. Unfortunately, when the proposal was debated by the Senate some members expressed serious misgivings about the amount, and rumours began to circulate that Stuart 'had made a good thing out of the workshop'. In the public discussion that ensued Stuart became embroiled in a damaging controversy over his request to the University to take over the workshop as a going commercial concern while remaining an essential part of the teaching of engineering. Eventually the whole enterprise was purchased by the University in 1887 on the understanding that Stuart would retain financial responsibility for the workshop, although its day-to-day operation would be entirely in the hands of Lyon, the Superintendent of the Workshops. Stuart agreed, not knowing that Lyon had no skill in the management of such an enterprise or in the keeping of accurate accounts.

Above: Madingley Hall, home of the University of Cambridge Institute of Continuing Education.

With the workshop controversy raging around him Stuart began a campaign to raise the status of the 'special' examination in Mechanism and Applied Science to the level of an honours degree – the Mechanical Sciences Tripos. Alternative schemes were put forward by opponents of engineering, but Stuart and his supporters believed that they had made a viable case for creating the Tripos. In February 1887, the proposal was put before the Senate, but to the surprise and dismay of Stuart and his supporters it was rejected by 73 votes to 59. Some members of the University privately held the view that the rejection was based less on the viability of the Tripos and more on a personal dislike of Stuart and his political activities – his attention having been diverted from University duties since being elected Member of Parliament for the constituency of Hackney in 1884. But this was not all: the radical views he had expressed at the hustings had outraged a predominantly conservative

RAISING THE ROOF: AN UNPARALLELED ACHIEVEMENT

Following a request from Stuart, approval was given for more space for his Department by adding a third storey to the nearby Department of Mineralogy. The intrepid engineers Stuart and Lyon, fearing that traditional building methods would cause long delays, devised a novel scheme for adding the extra storey. With careful design, the help of engineering students and under the expert supervision of Lyon, the existing 110-foot-long, 50-ton roof was jacked up and held in position while new walls were constructed underneath. The building was completed in just 17 days without any mishap, and 'not a single slate or nail in the roof was broken or strained'! The risky endeavour, initially opposed by the authorities, must have drawn crowds of spectators, and its success not only was highly praised but also greatly enhanced the status of the Engineering Department within the University.

Nineteenth-century view from the original Botanic Garden in Cambridge towards the first Engineering Department building.

University, and there was a stinging attack upon him in the press by one of his closest friends and former allies.

Worse was to come for Stuart. From 1887 to 1889 he paid little attention to his University duties, delegating them to the Superintendent of the Workshops, Lyon. This proved to be a disaster, with mismanagement in the operation of the workshop and negligence in the financial costings of projects. The fortunes of the workshop declined markedly and its poorly maintained accounts were heavily criticised. Although he had not been closely involved with the workshop for two years, Stuart took responsibility for it, and his weakened position allowed his opponents to question the very existence of the teaching of engineering. It was all too much for Stuart, and in December 1889 he wrote to the Vice-Chancellor that he wished to resign the Professorship of Mechanism and Applied Mechanics. His resignation was accepted and Stuart left Cambridge under something of a cloud, though we recognise Stuart today as the Founder of the Engineering Department, even if he did leave it resting on rather shaky foundations.

Stuart left Cambridge without achieving his aim of establishing the Mechanical Sciences Tripos. What then is the claim, suggested by his biographer Paul McHugh writing in *Cambridge Minds*, 'for inclusion in the company of Darwin, Keynes or Wittgenstein'? This exalted status rests principally on Stuart as a man of vivid imagination, immense capacity and energy in any endeavour, and his wide range of interests and accomplishments outside academia in radical politics, philanthropy and journalism. The range of his interests and his many achievements both within the University of Cambridge and in the outside world certainly place him among the greatest of the University's graduates, and the Engineering Department is justifiably proud of its Founder.

CHAPTER TWO

Laboratory and Tripos

James Alfred Ewing (1855–1935)

Professor of Mechanism and Applied Mechanics (1891–1903);
Head of Engineering (1891–1903)

The Professorship of Mechanism and Applied Mechanics was founded with the condition that it was 'to terminate with the tenure of the Professor first elected'. Thus, following the abrupt departure of Stuart, the incipient Engineering Department came under immediate threat. Although a University Committee's report had included the Professorship in the list of established chairs, hoping that this would tie the hands of the Senate and make the Professorship permanent, a minority of the Senate had remained virulent opponents of engineering and now sensed an opportunity to rid themselves of the subject altogether. A debate ensued between supporters and detractors, with neither side giving up its position. Fortunately, following pleas from the supporters of engineering, there was a welcome intervention by the Chancellor of the University, William Cavendish, Duke of Devonshire and founder of the Cavendish Laboratory. The Duke, an industrialist, was a committed supporter of the application of science in industry, and he decided unequivocally that the Professorship of Mechanism and Applied Mechanics should be not only renewed but established in perpetuity.

Following Stuart's resignation, a Syndicate enquiring into the future of the workshops had been appointed to make recommendations for the future development of the Department and, happily, it found that 'the continued existence of a school of engineering in Cambridge and its further development were highly important'. This report finally affirmed the teaching of engineering in the University and ensured that neither arguments based on the cost of teaching the subject nor those put forward on the unsuitability of practical subjects in the academic 'purity' of Cambridge scholarship would prevail in the future.

When the vacancy was announced, no fewer than 13 candidates presented themselves, including Alfred Ewing, then at the University College, Dundee.

James Alfred Ewing, second Professor of Mechanism and Applied Mechanics, consolidated the position of Cambridge engineering and established the Mechanical Sciences Tripos.

XXVIII May MDCCCCIII.

Above: *Turbinia*, the first steam-turbine-powered steamship in the world, designed by Charles Parsons and built in 1984, on display at the Discovery Centre, Newcastle upon Tyne.

Below: Ewing was Professor of Engineering at University College, Dundee, from 1883 until moving to Cambridge in 1891. In this 1889 photograph from Dundee he is fourth from the left.

Right: Letter of appreciation from staff and students presented to Ewing when he left the Engineering Department to become the Head of Education for the Admiralty in 1903. During the First World War Ewing and his team took on the task of deciphering coded messages.

Below right: Bust of John Hopkinson, Professor of Electrical Engineering at King's College, London. A close friend of Ewing, he was tragically killed in 1898 on a mountaineering trip in the Alps along with three of his children.

HYSTERESIS: WHERE DOES THE ENERGY GO?

While working in Japan Ewing investigated a phenomenon to which he gave the name hysteresis (from the Greek for 'lagging behind') and noted that hysteresis occurred in both mechanical and electrical systems. He studied the phenomenon experimentally in the cyclical magnetisation and demagnetisation of ferrous materials, demonstrating that a loop was formed during the cycle, with its area proportional to the work done per unit volume during the cycle; the result was fundamental to the understanding of power losses in electric transformers, motors and generators. Ewing also speculated on the generality of the phenomenon and over the years it has become apparent that hysteretic behaviour is seen in biological, medical and economic systems and is indeed a desirable property for system stability. His generalisation of the discovery is a model of clarity in scientific writing:

'When there are two qualities, M and N, such that cyclic variations of N cause cyclic variations of M, then if the changes of M lag behind those of N, we may say that there is hysteresis in the relation of M to N.'

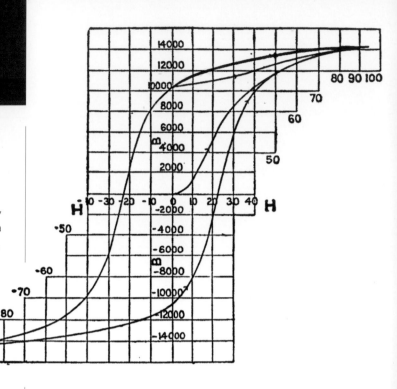

Above: Hysteresis demonstrated in this B-H loop of piano wire from Ewing's 1885 paper 'Experimental Researches in Magnetism'.

Below: Ewing was appointed Professor at the University of Tokyo at the age of 22 and is recognised in Japan for the contributions he made to the establishment of engineering in the country.

He was encouraged to apply by some of the more prominent supporters of engineering in the University. John Hopkinson, Professor of Electrical Engineering at King's College, London, was approached by Stuart to apply for the position himself but declined, and put forward Ewing instead.

On 12th November 1890, Ewing was appointed as the second Professor of Mechanism and Applied Mechanics. He was ideally suited for the position, not only because of his technical brilliance but because of the leadership he had demonstrated both in setting up the Engineering Department at the University of Tokyo in the early 1880s, and in establishing the School of Engineering at University College, Dundee. About this move, he wrote: 'From the bleaker regions of the Tay, Cambridge seemed a sort of Southern Paradise – a lotus eating land.'

A delay in confirming Ewing's appointment proved to be a stroke of luck for him. It was well into the Michaelmas Term before he was able to come to Cambridge to meet University officers and transfer his family, so it was agreed that he could take up his office at the beginning of the Lent Term, in mid-January. The time between his formal appointment and his arrival in Cambridge gave his supporters the opportunity to brief him thoroughly about the problems that had beset his predecessor, and of the bitter debate regarding 'the extent to which workshop practice should form part of the instruction

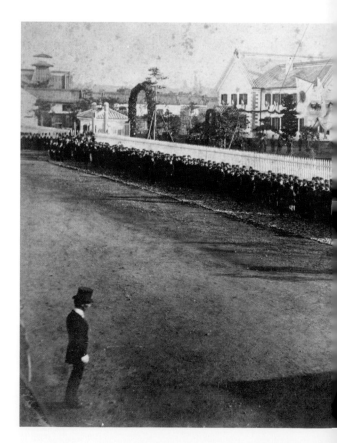

Fleeming Jenkin (1833–1885), Regius Professor of Engineering at the University of Edinburgh and the inventor of the cable car, inspired Ewing in his early studies. Jenkin's expertise was not limited to engineering; he was also an economist, linguist, artist, actor and critic.

at an ancient university'. He was told that the personal antagonism towards Stuart had been such that before his resignation the view had been expressed that 'Professor Stuart may be trusted to undertake no work in the Mechanical Workshops which would not enable him to teach and illustrate the subjects for the advancement of knowledge and study of which in the University the Professorship of Mechanism was founded'.

Ewing realised that his task in Cambridge would not be an easy one, and that he would have to make plans for his future course of action with some care. Visiting his inheritance he found just a handful of students, some of whom were interested simply in learning the use of tools, and there was no lecture room available to him until the eminent Sir George Stokes, Lucasian Professor of Mathematics, allowed him to use his lecture room when he was not himself lecturing. The room had very uncomfortable benches with neither back nor front, making note-taking a challenge.

Ewing was not one to equivocate and, within a few weeks of receiving a telegram from the Vice-Chancellor stating 'You are elected Professor', he laid out his aspirations for the Department in an inaugural lecture of January 1891 entitled 'The University Training of Engineers'. This was a defining moment in the Department's history. In the lecture Ewing informed his audience that he planned to establish a comprehensive School of Engineering that would be in sympathy with the sweeping recommendations proposed in 1850 by the Royal Commission. He also asserted that, in teaching the subject, there should be an initial general scientific study course followed by the specialised science courses needed for underpinning professional practice. Finally, he claimed that the well-trained engineer would require the 'acquisition of the practical knowledge which could only be gained by experience', but that no amount of practical instruction in industry without theoretical instruction in a university could make British engineers equal in quality to those already working in Europe. A newspaper account described his lecture as 'modest and resolute'.

The most important feature of his proposals was the pioneering proposition that a 'teaching laboratory' was an indispensable instrument of engineering education at any university, and he insisted that, as students would gain immeasurably from practice in experimental work before going on to work in industry, a well-equipped laboratory for measurements was essential even though it would be costly to equip. His request to the University was that he be provided with one laboratory for mechanical engineering and another for applied electricity, arguing that the existing workshops would be useful only as a part of such laboratory teaching. In his vision, workshop training would be limited compared with the practice followed in the time of his predecessor, and he proposed that the position of Superintendent of the Workshops should be replaced by a second demonstrator who would assist him with laboratory experiments. He ended his lecture with a forthright request to the University to

General Library of the University of Tokyo

The Engineering Department obtained its first substantial premises when it took over a building vacated by the Perse School on Free School Lane in 1890.

provide funds to meet his ambitions, although he had been told that the 'lid of the University chest was firmly shut'.

Whereas Stuart had become something of a *persona non grata* in the University towards the end of his tenure, Ewing was warmly welcomed as a colleague by the established professors in scientific disciplines, and shortly after his appointment he began to adopt a leadership role within the University.

The most pressing problem for Ewing was lack of space for his ambitious plans for a teaching laboratory. Fortuitously, in 1890, the Perse School for Boys in Free School Lane decided to move to a new site in Gonville Place, and the school buildings, which comprised a hall, a block of classrooms and two schoolmasters' houses, were purchased by the University. Ewing won a protracted battle against strong competition from other professors, acquiring the greater part of the available space. His requirements for a laboratory were as follows: 'It must hold a steam engine and boiler specially arranged for experimental work; tanks and other apparatus for hydraulic measurements; apparatus to experiment on the transmission of power; and a number of other bulky appliances. It should include rooms for dynamos and for experimental work in applied electricity.'

At the same time, he was able to fit out a lecture room for up to 120 students, a drawing office, research rooms, a woodworking shop, a fitting and turning shop for metalwork, an instrument shop, stores and a small foundry. The academic calm of the Museums Site was abruptly invaded by noisy steam engines and boilers, and there was a complaint that Ewing was disturbing the sleep of a valued horse in the stable of Edward Perowne, Master of Corpus Christi College, whose Lodge was just across Free School Lane. Noting the sudden riches bestowed upon Engineering, the humanities professors raised a cry that would be heard more than once – that the University had unfairly decided 'science first and the rest nowhere'.

Ewing's pleasure at his success in obtaining more space for his Department was compounded when a report from the Workshops Enquiry Syndicate appointed to review the position of engineering in the University stated that there was indeed a case for giving the subject a place similar to the one it held in other universities. The report stressed that the Professor of Mechanism and Applied Mechanics must be provided with 'those appliances which are now recognised as essential to the satisfactory treatment of the subject'. A leader in *The Times* read: 'Mechanical science is thus on the high-

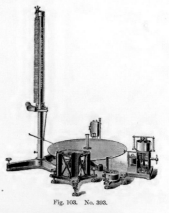

89

SEISMOGRAPHS.

These instruments were originally designed by Professor Ewing, F.R.S., for the Seismological Observatory of the University of Tokio. The forms now offered contain many improvements in detail, suggested by experience of earthquake measurement.

393. Horizontal and Vertical Motion Seismograph. The apparatus shown in Fig. 103 records two rectangular horizontal components and the vertical component of each successive displacement of the ground. The records are taken on a revolving plate of smoked glass which can be removed and replaced without disturbing the rest of the apparatus. A delicate seismoscope starts the plate revolving at the commencement of the shock and also starts a clock which beats seconds.

Fig. 103. No. 393.

Two spare plates are supplied with each seismograph.

Fig. 104 is an example of a small part of a record of an actual earthquake as given

road to much more hearty recognition and patronage than have ever before been accorded to it', but there was still some opposition, and a London evening newspaper muttered that 'before long our future Bishops would take first classes in brick-laying and paper-hanging.'

The University had contributed £20,000 towards the purchase of the site for Engineering but Ewing needed twice as much again for refurbishing the space and for equipping it to house the laboratory. The University, unable to provide these funds, launched a public appeal, but this raised only a fraction of the money needed. Apparatus for experiments could not be purchased from commercial suppliers and had to be manufactured somewhat crudely in the workshops, and Ewing had to seek gifts of equipment and books from sympathisers. Horace Darwin (founder of the Cambridge Scientific Instrument Company) and John Hopkinson were both very helpful, but were unable to provide the money Ewing needed. However, eventually the University's Financial Board made money available by appropriating funds from future development plans in the University. The new laboratories were opened in May 1894 by Lord Kelvin, with the Vice-Chancellor, the Provost of King's

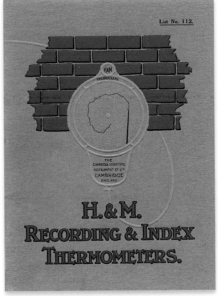

Above, left and opposite: Catalogues from the Cambridge Scientific Instrument Company Ltd, founded in 1881 by Horace Darwin (1851–1928) and Albert Dew-Smith (1848–1903). Darwin, son of Charles Darwin, was a civil engineer. Dew-Smith was an engineer and instrument maker. They were Cambridge graduates from Trinity College.

In 1874, Alexander Kennedy, at the age of 27, became Chair of Engineering at University College, London. A notable educationist, he reformed the teaching of engineering and constructed the first engineering laboratory in which students put theory into practice. He was an honoured guest at the opening of the Engineering Department's laboratories in 1894.

Charles Rolls (1877–1910) graduated in Mechanism and Applied Sciences in 1898 from Trinity College, Cambridge. His passion for cars led him to co-found Rolls-Royce Ltd with Henry Royce in 1906. Rolls was also an aviation pioneer and made the first non-stop double crossing of the English Channel by plane. He died at the age of 33 when his plane crashed.

An event in the Engineering Department at Free School Lane in 1907 with Lord Kelvin, second from the left on the platform, and on the right, Bertram Hopkinson.

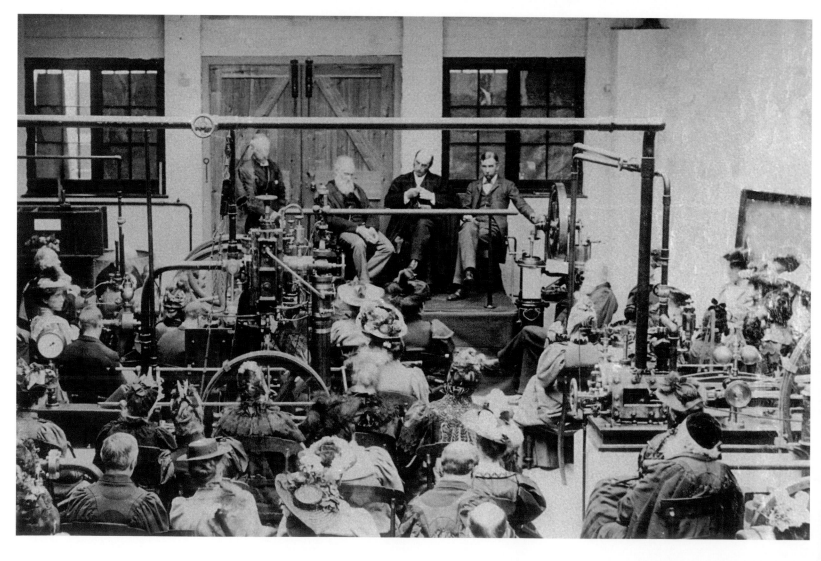

College, presiding. At the opening, a special tribute was paid to Alexander Kennedy, Professor of Mechanical Engineering at University College, London, who had invented the concept of the 'engineering laboratory' in which the subject was taught by experimentation and measurement. A moving spirit behind many of Ewing's ideas, he was an honoured guest at the opening ceremony, and rightly so, his ideas having stood the test of time.

THE MECHANICAL SCIENCES TRIPOS AND THE ENGINEERING DEPARTMENT'S EXPANSION

The Workshops Enquiry Syndicate reported that its members were 'unanimously of the opinion that the establishment of an Honours Examination specially adapted to the scientific training of students in the Mechanical Department is of urgent importance to the proper development of the study of the Mechanical Sciences in the University'. The Mechanical Sciences Tripos was thus created and the name remained in use until 1970.

Ewing concentrated on expanding his Department and accelerated its growth to such an extent that numbers grew from 74 undergraduates in 1894 to 226 by 1903, and staff members increased from two demonstrators to 11. Inevitably, this rate of growth (almost 10% of undergraduates in the University were reading Engineering in 1903) created an urgent need for more space. Fortunately, the University was also expanding and needed more land to build new premises, so it purchased six acres from Downing College, which later became known as the Downing Site, into which the Department of Botany was moved, with Engineering expanding into the space thus vacated. The University was unable to develop the site allocated to Engineering but the Hopkinson family's donation (see box overleaf) made possible the building of three well-equipped laboratories and a large lecture room, as well as some small rooms for research students. The new wing was opened in February 1900, again by Lord Kelvin in the presence of the Vice-Chancellor and a large assemblage, and was named the Hopkinson Wing after the late John Hopkinson and his son Jack.

The University recorded that during Ewing's 12 years as Head of Department the number of students had increased fivefold, while receipts from fees had increased by a factor of nearly 10, allowing the recruitment of no fewer than 14 new members of staff. Ewing's original demonstrators,

Ewing's 'two lieutenants' C G Lamb and J B Peace. They joined the Department as demonstrators in 1891 and in 1903 were promoted to University lectureships in Electrical Engineering and Mechanical Engineering respectively.

TRAGEDY IN THE ALPS: THE LOSS OF HOPKINSON

John Hopkinson, Professor of Electrical Engineering at King's College, London, a close friend and supporter of Ewing and a keen mountaineer, invited Ewing with his wife and children to join his family on a climbing holiday in Switzerland. In a dreadful accident on 27th August 1898, Hopkinson, along with his son Jack and his daughters Alice and Lina, was killed while climbing *La Petite Dent de Veisivi*. An accident had been feared when the party did not return in the evening as expected, and the next day all four bodies were found, roped together and lying 500 feet below the summit. Hopkinson was just 49 years old and in the prime of his illustrious career. The bodies were buried in Switzerland while the grieving widow, Evelyn, with her surviving child, Ellen, returned to England. (Another son, Bertram, was not in the holiday party.)

The surviving Hopkinsons, aware that the deceased John Hopkinson had been planning to assist Ewing with his aspirations for a new Engineering Department building, resolved to commemorate the lost members of their family by paying for the construction of a wing of the Engineering Laboratory in Free School Lane. His widow later made another donation towards the foundation of a Hopkinson Chair of Thermodynamics at the University.

In a footnote to this terrible event, following the death of Ewing's first wife in 1909, he maried Ellen, Hopkinson's surviving daughter.

C G Lamb and J B Peace, were promoted to University Lectureships in Electrical Engineering and Mechanical Engineering respectively.

During Ewing's term of office there was another important development within the University of Cambridge when the concept of postgraduate study was introduced, which in turn led to the expansion of research throughout the University. Graduates from other universities were attracted to Cambridge by courses of advanced study and opportunities for research. Some of these 'Advanced Students' were allowed to sit the Tripos after six terms of residence and admitted to the BA degree on successfully passing the examinations. Other Advanced Students were able to complete a year devoted entirely to research (precursors of the present-day research students) and receive a Certificate of Research; after a further three terms of residence these research students could also receive the BA degree. Cambridge graduates already in possession of a BA degree could also take the one-year course and obtain a Certificate of Research.

Where Stuart had founded the Engineering Department, Ewing had consolidated it and gained for Engineering the accolade of a Tripos equal in standing to the more established subjects in Cambridge. By 1903, the Department

Top: John Hopkinson (1849–1898).

Above: Ellen Hopkinson, who married Alfred Ewing.

of Engineering was held in high esteem throughout the country and Ewing had become a prominent national figure, receiving compliments from across the nation. His most remarkable contribution was that he had successfully fought against the prejudice 'that it would be well for Cambridge to confine its energies to the promotion of the older studies, leaving those which had a direct bearing on commerce and industry to be dealt with in universities then springing up in the main centres of industrial life'. Much later, in one of his lectures he quoted Lord Kelvin: 'There cannot be a greater mistake than looking superciliously on the practical applications of science, the life and soul of science is its practical application.' In 1903, Ewing suddenly changed path, resigning his position and leaving Cambridge to pursue a career in the service of the nation, making contributions of great value to naval intelligence and code breaking.

Above: Plaques commemorating John Hopkinson, Professor of Electrical Engineering at King's College, London, and his son, who died tragically in the Swiss Alps in 1898. A donation from the remaining family members enabled the Engineering Department to open a new wing in 1900. Hopkinson's son Bertram succeeded Ewing as Head of Department.

Left: The practice of drawing a pig while blindfolded was a popular game in Ewing's time. In 1899 *The Strand* magazine published a collection of blindfolded pig drawings by 'leading representatives of science, art, literature and society'. Soon, the craze spread to America as this advertisement from 1906 explains: 'The Pig Book is the novelty of the year and Pig Parties are everywhere. There is no fun equal to that of watching each blindfolded guest draw a pig in the Pig Book.' Ewing was a huge fan and compiled pig sketches by his distinguished guests.

The page on the left from Ewing's book includes pictures by Margaret Sackville, H P Macmillan, Siegfried Sassoon, Stanley Baldwin and Winston Churchill.

CHARLES PARSONS

Frequently cited as the greatest British engineer since James Watt, Charles Parsons (1854–1931) studied mathematics at St John's College, Cambridge and attended Stuart's lectures on mechanism and applied mechanics. His studies in Cambridge and also Trinity College Dublin gave him the academic foundation to revolutionise power generation with his novel steam turbines and create a new era in marine engineering. Among his inventions were a four-cylinder high-speed epicycloidal steam engine and a high-speed dynamo that was coupled directly to a steam turbine. His turbine used the principle of subdividing the expansion of steam into a number of stages with pressure drops between the stages, thus reducing the velocity of the impulse on the turbine blades to a manageable level, with both the action and reaction of the steam utilised to drive the blades. In 1884 he built his first multistage reaction turbine, which developed an output power of 7.5 kW running at a speed of 18,000 rpm – a revolutionary breakthrough. The essence of Parsons' patents had come from an intuitive moment of genius when he sketched the design of the reaction blades on the back of an envelope, and this design became the industry standard; it took years of research and the expenditure of large sums of money to improve the efficiency of the design by a few per cent.

Parsons' first customer, the Forth Banks power station at Gateshead near Newcastle, installed two Parsons 75 kW turbo alternators to provide power for the public lighting in the city, the first application of the invention. This was closely followed by the installation of three four-ton, 100 kW turbo alternators in Cambridge in partnership with Ewing, then Head of the Engineering Department. In time, as turbine alternators increased in efficiency and capacity and decreased in size as a result of technical progress, steam engines were replaced by Parsons' machines in power stations across the world.

In his original patent Parsons had considered the use of steam turbines for ship propulsion, but several years elapsed before he had the time to establish another company, the Parsons Marine Steam Turbine Company, for the purpose of exploiting his patent in marine applications. To test his

Parsons revolutionised the generation of power with his invention of the modern steam turbine.

Parsons' steam turbine generator from 1884.

ideas, he built a small vessel, *Turbinia*, about 100 feet long and with a particularly slender hull compared with conventional ships. Initial trials of *Turbinia* proved to be disappointing – its maximum speed was just 24 knots. Parsons investigated and discovered that a single propeller could not produce the expected thrust from his powerful turbine no matter how fast the blades turned; *Turbinia* needed multiple shafts each with multiple propellers to increase the total surface area of the blades and reduce their speed. These changes enabled *Turbinia* to reach a speed of 34 knots – greater than the speed of the fastest naval ships at that time.

Parsons, however, did not rest with that spectacular result. His experiments continued in order to identify and fully understand why *Turbinia* had not been able to go faster with a single propeller. The answer lay in the formation of efficiency-sapping cavitation bubbles. His work in this field, which continued to the end of his life, provided the foundation for not only avoiding the harmful effects of cavitation in propulsion systems, but also exploiting the effect in chemical process industries.

Turbinia was built by Parsons as an experimental vessel in 1894 to demonstrate the first use of the steam turbine in marine propulsion. Its astounding performance made it the fastest ship in the world at that time, which was demonstrated dramatically at the Spithead Navy Review in 1897. *Turbinia* set the standard for the next generation of steamships. Ewing was a close friend of Parsons and collaborated with him on the development of the steam turbine, taking part in the sea trials of *Turbinia*.

CHAPTER THREE

Raising Standards

Bertram Hopkinson (1874–1918)

Professor of Mechanism and Applied Mechanics (1903–1914);
Head of Engineering (1903–1914)

Following the unexpected resignation of Ewing, there was an interregnum of some months during which the administration of the Department became the responsibility of J B Peace, Lecturer in Mechanical Engineering, who had served in the Department for many years under Ewing. Expecting that Ewing would support his application for the Chair, he put himself forward as a candidate, but instead Ewing encouraged his brother-in-law, Bertram Hopkinson, to apply. Ewing made a strong case in support of Hopkinson, emphasising his exceptional personal qualities and matching them to the perceived needs of the Engineering Department, and Hopkinson was duly appointed Professor of Mechanism and Applied Mechanics at the age of 29.

Interested parties, aware that Bertram was the son of Ewing's friend and mentor John Hopkinson and that his sister Ellen was Ewing's second wife, may have looked askance at the hint of nepotism in the election of so young a professor with a very short career in engineering who was untried in teaching. But Hopkinson's tenure in office was to prove that any misgivings were wholly unjustified, as he turned out to be an exceptional leader. Ewing described him as 'tall, of very commanding presence, with immense physical strength and energy, with ripe engineering experience and great originality of mind he commanded respect and confidence in all those who worked with him.'

Hopkinson was on track to become a barrister in 1897 after graduating from Trinity College in Mathematics when he was devastated by the news of the death of his father and three of his siblings. He immediately changed the direction of his career and joined the family's engineering consultancy, and quickly established a considerable reputation as a scientist–engineer capable of applying mathematical and physical analysis to practical engineering problems. In 1903, he married Mariana Dulce Siemens, eldest daughter of

Bertram Hopkinson, third Professor of Mechanism and Applied Mechanics, raised the standard of engineering teaching in the Department and led a remarkable range of research activities that influence engineering to this day.

Above: Departmental photograph from 1906. Hopkinson is seated in the middle of the front row; on his right is Charles Inglis, who went on to become Head of Department.

Above right: During the First World War Hopkinson dedicated himself to war work. He qualified as a pilot and was killed in a flying accident in 1918.

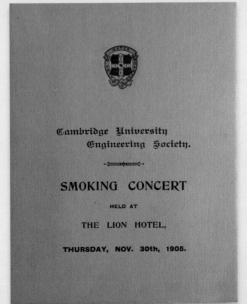

Above: A car designed by Harry Ricardo fitted with his Dolphin engine. Ricardo developed the first Dolphin engine while he was an undergraduate: 'I then designed and built, also in the workshops at Cambridge, an improved version, which was completed in 1905 and was put through fairly thorough calibration tests in the Engineering Laboratory.'

Above right: Ricardo's two-stroke Dolphin engine fitted up for brake testing.

Right: An early programme for an event organised by the Cambridge University Engineering Society.

Cambridge University
Engineering Society.

SMOKING CONCERT

HELD AT

THE LION HOTEL,

THURSDAY, NOV. 30th, 1905.

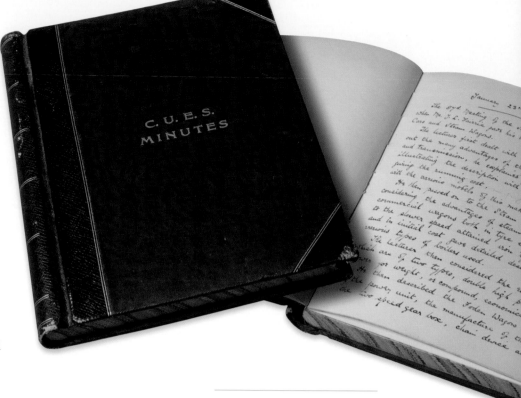

Alexander Siemens and a cousin of the founders of the famous German firm of electrical engineers, and in the same year he was elected to the Chair of Mechanism and Applied Mechanics at the University of Cambridge.

At the time of his joining, the Engineering course was structured in two streams: the Honours degree (the Mechanical Sciences Tripos) and an Ordinary Degree. However, the standard of both degrees was questionable, and Hopkinson was dissatisfied with his inheritance. Aware that he had no experience of academic institutions, he proceeded with caution. His first action, delayed to 1905, was an attempt to raise the standard of the Tripos by demanding more mathematical skills in men wishing to study engineering, through the introduction of a Qualifying Examination.

Acting in the wider sphere of the University, he supported and even led the campaign for the revision of the Mathematics Tripos, against bitter opposition from the private coaches who taught mathematics to undergraduates following a traditional syllabus. He proposed that Part I of the Mathematics Tripos should be an elementary examination at the end of the first year, suitable as a qualification not only for mathematicians wishing to proceed to the more rigorous Part II of the Tripos but also for students wishing to transfer to Engineering, contending that 'the science of engineering must be the product of a union between mathematics and physics, and it was the mathematical engineer which Cambridge ought to produce'. New regulations came into effect for the Mathematics Tripos in 1906 based on the ideas put forward by Hopkinson and other reformers.

Initial fears that the higher standards demanded by Hopkinson would deter students from reading Engineering proved groundless: there was only a very slight drop in applications for a year or two, which was compensated for by the higher standards of those applying, and the Honours Degree course gained in reputation. Under Hopkinson's leadership the Department went from strength to strength on all fronts, with student numbers growing to 260 by 1910. Engineering graduates from Cambridge found ready employment, often rising rapidly to senior positions of management in industry. Hopkinson also

The Cambridge University Engineering Society (CUES) was founded in 1901. Shown here is a minute book from 1905. Today CUES is the largest academic society in Cambridge with more than 1,000 members among current students, and more than 20 times more alumni members worldwide.

PREPARING PRACTICAL ENGINEERS: THE POLL MEN

'Poll men' was the name given to those who were reading for the Ordinary Degree at the University of Cambridge, and their needs were frequently overlooked while the standard of the Tripos was raised. But Hopkinson was very aware of them, and he introduced new regulations for the Special Examination suitable for their abilities, insisting that poll men must be well taught, with adequate departmental resources devoted to their needs. He believed that men who graduated with the Ordinary Degree often did extraordinarily well as practical engineers, not only in the UK but also abroad. In 1913, the Special Examination in Mechanism and Applied Science was renamed the Special Examination in Engineering Science, but the title of the Mechanical Sciences Tripos was left unchanged. It would take another 60 years for the name to be changed to the Engineering Tripos.

took a personal interest in laboratory teaching, and meticulously inspected the undergraduates' experimental laboratory notebooks, discussing each experiment with the student in some detail before approving his work. A strict and demanding teacher, he was held in great awe by all who encountered him.

The Department staff comprised two lecturers, Peace and Lamb, two demonstrators, including Charles Inglis, who was destined to be Hopkinson's successor, and 10 assistant demonstrators. The disappointed Peace left the Engineering Department in 1908 to take up the position of University Printer in the Cambridge University Press, and Inglis was promoted to the Lectureship in Mechanical Sciences vacated by him. The course principally comprised papers in mathematics, mechanics, strength of materials, theory of structures, heat, heat engines, electricity and magnetism. Academic staff members were expected to teach 'across the board', in other words with equal facility in all subjects – a condition which effectively debarred those who had graduated from universities other than Cambridge from appointments in the Engineering Department. This requirement for polymathic teaching at the Department remained in force until well after the end of the Second World War.

Hopkinson became aware that the workshop was being operated in a slipshod manner, with lax workmen coming in at all hours and no check being kept on either their timekeeping or their daily attendance. Students were in the habit of interacting directly with the workmen for their construction projects and from time to time students' practical assignments were carried out surreptitiously by the workmen in exchange for

A pensive Hopkinson captured at work in his study. Ewing wrote after Hopkinson's premature death that he had a great 'originality of mind' and with his 'unruffled kindliness and serenity, he commanded respect and confidence in all those who worked with him'.

HARRY RICARDO

The internal combustion engine was still a novelty when Harry Ricardo (1885–1974) began his career, but he was fascinated by engine technology and developed a passion for understanding its underlying physical principles. His seminal contributions to internal combustion technology, spanning half a century, helped to make the engine ubiquitous before his death in 1974. The physics of ignition, the principles of combustion and detonation, in both petrol and diesel engines, and the concept of octane ratings for fuels were just some of the subjects of his research. He designed four-stroke engines for armoured tanks used to great effect in the battles of the First World War, as well as diesel engines for trucks, tractors and taxis, and helped to develop the sleeve valve engine for aircraft, placing him amongst the greatest of all British engineers.

Ricardo designed and built his two-stroke Dolphin engine in the workshops during his last year at Cambridge, and later founded a small firm to take the design into production. The engine became popular as a marine engine for fishing boats and was also fitted into a range of cars under the name of 'Dolphin', but the venture was a commercial failure and abandoned in 1911. Undeterred, in February 1915 he registered another company, Engine Patents, now Ricardo plc, which continues to this day to focus on creating new technologies and innovations for manufacturers engaged in the mass production of engines. Today, the company he created has been in business for over 100 years, with more than 2,000 employees and a turnover of more than £200 million.

Ricardo, one of the foremost engine designers and researchers in the early years of the development of the internal combustion engine.

In 1915 Ricardo set up a company that developed the engine for the Mark V tank.

sums of money from the 'young gentlemen'. Hopkinson, not one to tolerate any improper practices, fixed the timekeeping problem by standing with his commanding presence in front of the workshop at precisely eight o'clock in the morning, adopting an unforgiving demeanour and holding his timepiece prominently in his hand; timekeeping promptly improved. At the same time, G F C (Freddie) Gordon of Trinity College was appointed to a supervisory role in the workshops, with instructions to demand a higher level of performance from the staff and put an end to all corrupt practices.

ESTABLISHING RESEARCH IN THE ENGINEERING DEPARTMENT

Hopkinson had inherited a passion for research from his father and his aim was to make the Engineering Laboratory a centre of research that would compare favourably with the Cavendish Laboratory – no mean ambition, as the physicists were already winning Nobel Prizes. Unlike his predecessors, he did not have to fight for space for the laboratory or to concern himself with serving on University committees, and he had an adequate number of academic staff to assist him with teaching duties, which had been an enormous burden on Stuart and Ewing.

EXPLOSIVE IDEAS: THE HOPKINSON BAR

In September 2014, a meeting was held in Cambridge to commemorate the centenary of the publication of a paper by Hopkinson in the *Philosophical Transactions of the Royal Society*. In this paper, he had described 'a method of measuring the pressure produced in the detonation of high explosives or by the impact of bullets'. The originality and quality of the work was remarkable, and it has withstood the test of time. He devised the apparatus that came to be known as the Hopkinson pressure bar, in which he used a ballistic pendulum 'to determine the momentum trapped in a short steel rod of known length magnetically attached to the far end of a long steel rod when the distant end is subjected either to impact by a lead bullet or to the blast produced by the detonation of gun cotton'. The longitudinal blow to the rod is transmitted as a compressive wave through the short bar, but when it is reflected at the far end of the bar it returns as a wave of tension, causing the magnetically attached short piece to be detached from the longer bar. By varying the length of the short piece, the full momentum–time profile of the blow could be determined. This elegant experiment and its variations have been used by research workers for 100 years.

With one group of research students, Hopkinson worked on the strength of metals under explosive loading, while further groups researched with him on fatigue hysteresis in metals subjected to cycles of stress, on the magnetic properties of iron and its alloys and on the internal combustion engine. His personal prowess and grasp of fundamentals across the whole spectrum of early twentieth-century engineering was quite astounding.

The most famous of his students, Harry Ricardo, who made a name for himself with his pioneering work on the internal combustion engine, sang Hopkinson's praises as a research supervisor 40 years after he had himself graduated from the Engineering Department. He said of Hopkinson that he was the 'most brilliant,

Hopkinson's pressure bar has stood the test of time and remains a standard piece of equipment in use in laboratories around the world.

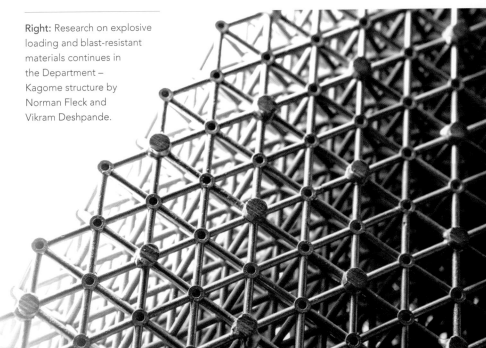

Right: Research on explosive loading and blast-resistant materials continues in the Department – Kagome structure by Norman Fleck and Vikram Deshpande.

The number of students plummeted during the First World War. This photograph shows how depleted the third year had become in May 1915. C G Lamb (front row wearing a hat) led the teaching. A H Peake, one of the demonstrators, is sitting on his left next to graduating student E L M Keary of Newnham College, who went on to publish papers on aeronautics, steering and screw immersion.

versatile and imaginative research leader I have ever met'. Under Hopkinson's leadership the Department became a significant centre of engineering research.

Ewing wrote that '[Bertram] Hopkinson was himself devoted to research and could inspire his pupils with a like ardour,' and compared Bertram's publications with those of his father John, noting that 'there is something of the same freshness of outlook, the same penetration and grasp, the same personal detachment, the same directness in attack, the same unconventionality in method, the same avoidance of side issues and concentration on the essence of the problem'. Before he left the Engineering Department, he had established not only a fine teaching department but also a centre of scholarship and research. One of his inventions, the Hopkinson–Tring torsion meter, is still used to measure the power delivered by the shaft in a turbine-powered ship. In 1910, Hopkinson was elected a Fellow of the Royal Society and just before the war a Professorial Fellow of King's College.

The onset of the First World War meant that numbers within the Engineering Department were rapidly depleted. From hundreds of engineering undergraduates in 1914, the number had fallen to just 17 in 1916. Perhaps the most significant among the exodus was Hopkinson himself, who transferred all his activities into war work, serving with distinction for four years and throwing all his energy into the war effort. In Hopkinson's absence, Lamb and T Peel undertook the teaching of the handful of remaining students with Peace supervising the workshops.

Although considered 'too old to fly' (he was twice the age of wartime pilots), Hopkinson qualified as a pilot and worked on the problems of flying at night, in bad weather and while navigating in cloudy conditions. In the final year of the war, and somewhat belatedly, he was invested Companion of Saint Michael and St George, promoted to the rank of Colonel and appointed Deputy Controller of the Technical Department of the Air Force. In the same year he was mentioned twice in the Secretary of State's list for 'valuable services'.

Hopkinson was killed in 1918 when his Bristol fighter crashed during a solo flight. Flying over London, he encountered difficult conditions in low cloud and lost control, crashing at High Ongar in Essex. He was given a hero's burial at a service in Cambridge, with tributes from his many admirers. J J Thomson, Master of Trinity College, President of the Royal Society, Nobel Laureate, said of him: 'Our Roll of Honour contains the name of no one who has rendered greater service to his Country.'

CHAPTER FOUR

A New Home

Charles Inglis (1875–1952)

Professor of Mechanism and Applied Mechanics (1919–1934);

Professor of Mechanical Sciences (1934–1940); Head of Engineering (1919–1943)

The untimely death of Bertram Hopkinson created an immediate vacancy for the headship of the Engineering Department but, because of the exigencies of war, there was a delay of some months before the electors to the Chair could convene a meeting. Meanwhile Charles Inglis, the best qualified and most experienced pre-war staff member, returned from wartime duties to his post of Lecturer in Mechanical Engineering. The electors, mindful of his teaching prowess and of his wartime engineering achievements, which included the design of several strategically important bridges for the military, notably the Inglis Pyramid Bridge and the Inglis Heavy Type Bridge, appointed him Head of Department. Ewing again played a part in this appointment: as one of the electors, he recalled that Inglis was 'a product of his time'.

The admission of undergraduates to the University had been all but suspended in 1914 and expenditure on the maintenance of the Department had been limited to a bare minimum during the four war years; equipment had deteriorated and not been replaced. Some staff members had been killed or wounded in action, others had retired and suitable replacements had not been forthcoming. Many of the young men destined to come up to Cambridge over the war years had been killed or wounded in action while some of those who had survived were unable to seek or resume a university education, being obliged to take up paid employment to support their families. Thousands of potential engineering undergraduates were lost – a sad legacy of the years of war.

Unlike his predecessors, all of whom were external appointments, Inglis had worked under Ewing as an assistant demonstrator and under Hopkinson as a demonstrator and later a lecturer, and was therefore entirely familiar with the functioning of the Department. Early in his career he had been recognised as an outstanding teacher and he had carried the largest teaching

Charles Inglis, notable civil engineer and Head of the Engineering Department, who oversaw the Department's move from Free School Lane to the Scroope House Estate.

Right: Extract from student's notebook, 1924.

Below: Students with steam engine at Free School Lane, 1920.

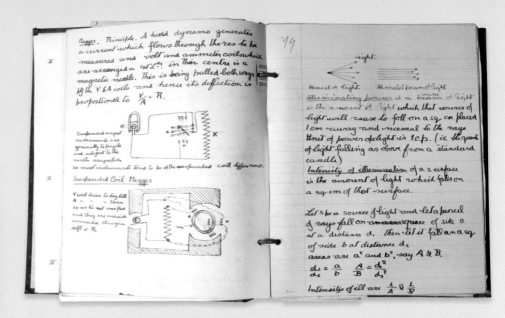

Below: Students in the Drawing Office in the early 1920s (now part of the ground floor Hydraulics labs in the Inglis Building). The three female students on the left, wearing hats, were reading Mechanical Sciences.

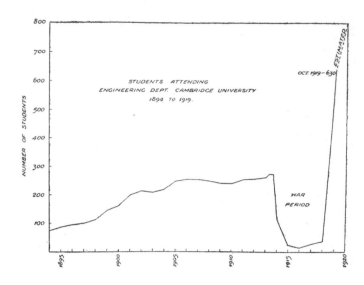

load in the Department, covering an unusually wide range of engineering disciplines. His experience was invaluable, given the overwhelming task of the reconstruction of a Department that had been virtually moribund during the war years.

In the 'peace years' there was a rapid increase in academic staff and student numbers throughout the University and many improvements were made in the facilities for research in the arts and in the natural sciences. The Cavendish Laboratory gained great fame for its physics research, while Inglis concentrated on teaching rather than research in the Engineering Department. During these 21 years of peace, many buildings were erected and the University of Cambridge gained great prestige internationally. New laboratories, museums and libraries were established and it was said of this time that 'a "progeny of golden years" was permitted to descend and bless us'.

At the start of the 1918 academic year there were just 37 students in the Department. A year later there were 553 registered for the Mechanical Sciences Tripos as well as a number of naval officers who had been seconded to take special courses in engineering; similarly, the teaching staff, which had numbered just two during the war years, rose to 38. Of these, 11 had returned from the war and the rest had been energetically recruited by Inglis against fierce competition from other prospective employers. Gordon was reappointed Superintendent of the Workshops and L C P Thring was appointed Superintendent of the Drawing Office. By the Michaelmas Term of 1920, there were 48 academic staff members teaching a record total of 808 students, compared with the pre-war total of around 200.

Above: Graph showing the decline in student numbers during the First World War and the sharp increase in numbers after the war.

Below left: During the First World War, Inglis served in the Royal Engineers. He designed a reusable steel bridging system known as the Inglis Bridge which was adopted by the military. Shown here is the Inglis Mark I which was used widely in the final year of the war.

The front of the first CUES *Journal*, published in 1921.

Above: Students perched on the roof (left) and the Head of the Workshops with an engine (right) at Free School Lane, 1920.

Below: Main workshop at Free School Lane.

Such numbers could not be accommodated in the existing buildings, and army huts had to be hastily erected on the tennis courts adjacent to the Department for use as lecture rooms and laboratories. Fortunately for Inglis, new regulations enabled men with military experience to graduate in two years instead of three, and rules of residence were relaxed; those who could not be accommodated in college rooms were allowed to go into lodgings in the city. An insurmountable difficulty was the dire shortage of laboratory apparatus, not just in Engineering but across the whole of the University, and Inglis had to improvise by building apparatus in the workshops to provide for his teaching needs. Some of this laboratory equipment was still in use in the 1960s, still perfectly serviceable and entirely suitable.

During the war years, academic staff members had been very poorly paid – not even a living wage in some cases. After the war, the government decided to give each university a single block grant to be distributed to all departments. The University Grants Committee was set up to advise on the sum to be allocated in each case. The University of Cambridge sought a substantial government grant, but remained fearful that its long history of freedom from political influence might be compromised. When the grant was received there were indeed conditions attached to it: a Royal Commission was appointed to look into the relationship between the colleges and the University, which decided that Oxford and Cambridge universities would, in future, receive an annual grant-in-aid from the Treasury, and that a system of graduated taxation of the colleges would be instituted. The more welcome announcements were that the salaries of University staff would be considerably enhanced and that there was to be provision of studentships for graduates to enable them to pursue advanced studies. It is of particular interest to note the proposition for separate funds to be allocated for women's university education and for extramural work.

SCROOPE TERRACE ON TRUMPINGTON STREET

The post-war financial straits extended to the Engineering Department, where the cramped conditions in the New Museums site had become intolerable as student numbers increased, and a new site was urgently sought. A property off Trumpington Street, known as the Scroope House Estate, with about three acres of land, was for sale, and although some Department members thought the site was too remote from their colleges (a little over 700 metres from the existing laboratory in Free School Lane), the University purchased the land for the Engineering Department for a sum of £14,900. The owners of the site, Caius College, persuaded the sitting tenant, Dr Wingate, to give up his estate, rather reluctantly, with several years of his lease still to run, for the good of the University.

Inglis wrote: 'The site has the further advantage of being screened behind a row of buildings in Trumpington Street known as Scroope Terrace, so that for the time being the new buildings can be simple in character and no money

Aerial photograph of the Scroope House site and surroundings from the early 1920s. (Scroope House is in the middle.)

need be uselessly expended in architectural elaboration.' Inglis immediately approached the architect F W Troup, known to have had long experience in the layout of modern factories, to draw up plans for a purpose-designed building for his Department. Troup's concept was a rectangular building measuring 300 feet by 350 feet with a frontage towards Trumpington Street and backing on to Coe Fen, with three main laboratories, several lecture rooms, a workshop and a drawing office.

However, Inglis was disappointed to learn that after buying the site there was no money in the University's coffers to pay for the erection of the building. He had hoped for more and wrote: 'The [Department] buildings are nevertheless planned with an eye for the future and with the idea that at some more or less distant date, when the Engineering Department may wish to blossom into super-magnificence, Scroope Terrace may disappear and a frontage thereby be obtained on Trumpington Street.' Today, although super-magnificence has indeed come, Scroope Terrace remains.

An Appeal Committee was set up but very little money was available by way of donations from a public impoverished by the war, and Inglis feared that it would take a decade or more to raise the necessary funds. Happily, there was an unexpected and remarkable intervention. The Vice-Chancellor received an unsolicited letter from Sir Dorabji Tata, a graduate of Caius College and proprietor of the Tata Engineering Company of India, with a generous offer of help.

Plaque commemorating the donation from the industrialist and former member of the Department, Sir Dorabji Tata, which covered the cost of the first stage of the building on the Scroope House site.

Left: The Electrical Laboratory. The date is unknown but sitting at the front is C G Lamb, who retired in 1934.

THE FRANCIS MOND PROFESSORSHIP OF AERONAUTICAL ENGINEERING AT THE UNIVERSITY OF CAMBRIDGE

The Francis Mond Professorship of Aeronautical Engineering, the first endowed aeronautics chair at a British university, was created as a result of a gift from Emile Mond, a prominent industrialist and philanthropist. Mond was moved to make this donation after his son Francis, a Peterhouse alumnus, was killed in 1918 while serving with the RAF in France. Mond stipulated that the University should support the appointment by providing working space, equipment and other resources for a branch of engineering that would be of ever-increasing national importance, and the Air Ministry agreed to situate an experimental station at an aerodrome near Cambridge. The Professorship was initially assigned to the Special Board of Mathematics, but was soon reassigned to Engineering.

The Professorship has been held by:

• Bennett Melvill Jones, 1919–1952

• William Austyn Mair, 1952–1983

• Michael Gaster, 1986–1995

• Bill Dawes, 1996–

Sir Dorabji's gift of £25,000 (about £10 million in 2016), together with the sum raised by the Appeal Committee, made it possible for the greater part of the building designed by Troup to be erected and made ready for partial occupation by December 1922. The original building on the site, Scroope House, became the administrative office for the Engineering Department, and the new Department was fully operational by the Michaelmas Term of 1923. GEC presented valuable electrical equipment, and many other engineering companies also provided equipment for the laboratories. Inglis wrote: 'The Engineering Department will be thoroughly efficient, up-to-date, and worthy of the high-class material which in ever increasing excellence comes to Cambridge to receive its engineering education.'

Following its transfer to the Scroope House Estate, the Department became the largest in the University, with approximately 500 engineers in residence each year. New University statutes were introduced in 1926 and the Regent House was instituted to enable University business to be conducted with the assistance of professional administrators, making it more efficient than hitherto. A Faculty Board

Above: Francis Mond, in whose memory the first Chair of Aeronautical Engineering at Cambridge was endowed.

Below: The Inglis Building, as it is now known, as it stands today.

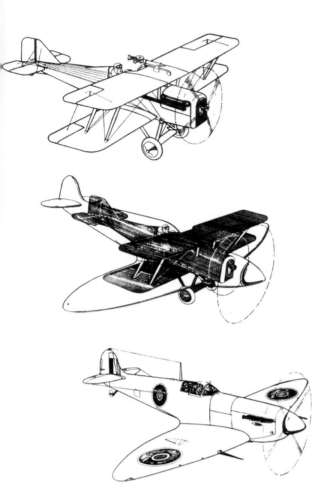

Bennett Melvill Jones (1887–1975) pioneered the streamlining of aeroplane design. His research led to a radical change in design, with braced biplanes, used during the First World War, being replaced with the clean monoplane, typified by the Spitfire launched in 1938.

of Engineering was established and stipends of University lecturers and demonstrators were increased substantially.

The tenured academic staff at the Department comprised the Professor of Mechanism and Applied Mechanics, the Reader in Electrical Engineering (Lamb), the Lecturers in Thermodynamics and Mechanical Engineering, three demonstrators and the Superintendents of the Workshop and the Drawing Office; the remaining staff members were without tenure. The range of subjects covered for the Tripos now included all the main branches of engineering – civil, mechanical, electrical and aeronautical – and Inglis concentrated his and all his staff's attention on teaching, to raise the standard to a very high level. In 1934 Inglis' Professorship of Mechanism and Applied Mechanics was renamed the Professorship of Mechanical Sciences, in line with the title of the Tripos.

AERONAUTICAL ENGINEERING

Bennett Melvill Jones was elected the first Francis Mond Professor of Aeronautical Engineering in 1919 and Inglis, with commendable foresight, offered accommodation and secretarial help to Jones, but the only room available for the Professor's office was in the former servants' quarters of Scroope House. An invitation was extended by Inglis to the new Professor to collaborate with the Engineering Department, which was warmly accepted, and an examination paper on aeronautics was added to the subjects available for the Mechanical Sciences Tripos. Inglis' foresight in establishing a good working relationship with Jones was rewarded in 1932 when the Aeronautics Department formally became a sub-department of the Engineering Department.

With no immediate prospect of obtaining a building from the University, Jones temporarily borrowed a wooden hut in the grounds of the University Air Squadron for his research – and then stayed there for 33 years until his retirement in 1952. A member of Jones' team, William Farren, designed two wind tunnels: the first in 1928 and the second in 1935. They were built in the workshops of the Engineering Department and installed in the wooden hut; one of these wind tunnels was transferred to the Engineering Department in 1959, when space did become available for Jones' successor, Austyn Mair.

Groundbreaking research was carried out in the Jones Wooden Hut Laboratory and twentieth-century aerodynamics owes a great deal to this pioneering work. Although he experimented with models in wind tunnels, Jones had strong views on the need, in aeronautical research, for data to be gathered while flying an aircraft. He wrote in 1930: 'It ought to be realised that free flying experiment provides the essential cutting-edge aeronautical research.' Farren, speaking on the occasion of Jones' retirement, said: 'He has created a laboratory both in the Engineering School, and in the air.'

49

Jones was the first to demonstrate the importance of streamlining in the design of aircraft and the remarkable benefits of ensuring that the aerodynamic drag on an aeroplane is limited to the skin friction on such surfaces of the aircraft as are essential for it to fly. Jones presented his seminal work, 'The Streamline Aeroplane', to the Royal Aeronautical Society in 1929 and received immediate and unstinting praise from informed members of the audience when he demonstrated the large gain in speed that could be obtained without any increase in power by using a streamlined design. The external form of an aeroplane, deliberately designed so that air could flow smoothly over its surface without turbulence in the thin boundary, was proposed by Jones. The effect on aircraft design was immense, and led to 'the introduction of the retractable undercarriage, hitherto considered to be too heavy and complicated, and the change from the braced biplane to the clean monoplane'. Jones and his team continued to work for more than two decades on reducing drag, laying the foundation for subsequent research by others on the design of modern aircraft.

In another aspect of his research, Jones studied the highly dangerous stalling of an aircraft when it reached a speed near the lower limit for steady flight in flight manoeuvres. The work, presented as the Wilbur Wright Lecture of the Royal Aeronautical Society and entitled 'Stalling', enthralled the audience as Jones presented a complete understanding of, but no cure for, the problem. His explanations led to significant improvements in design that almost eradicated stalling catastrophes. Remarkably, his team of research workers was always a small one, with just three or four members, and he maintained that he expected his team to work with him and not for him.

Portrait of Melvill Jones, first Francis Mond Professor of Aeronautical Engineering, which is currently displayed in the Aerodynamics Laboratory.

Front cover and diagrams from a report on stalling in aircraft conducted by Jones for the Ministry of Defence in the early 1930s.

RESEARCH AND TEACHING UNDER INGLIS

In his time, Inglis published numerous scientific papers based on his personal research and gave highly commended public lectures, not only on his research interests but also on more general topics in education and engineering. He did not, however, favour a research-oriented Department, and he disliked specialisation, expressing the view that 'premature specialisation cramps the imagination and it is destructive to the length and breadth of mental vision'. In a university setting he argued for the teaching of engineering principles rather than the day-to-day practice of engineering, and wrote: 'In a University course of engineering, instruction should primarily concentrate on teaching those essentials which, if not acquired at that stage, will never be acquired. The technicalities which

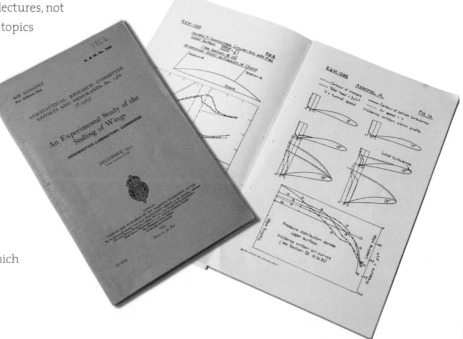

The Inglis Building (seen from Coe Fen) was expanded in the early 1930s to include well-designed lecture theatres, still in use today, laboratory teaching facilities and a workshop.

Layout of the ground floor of the Inglis Building, 1933.

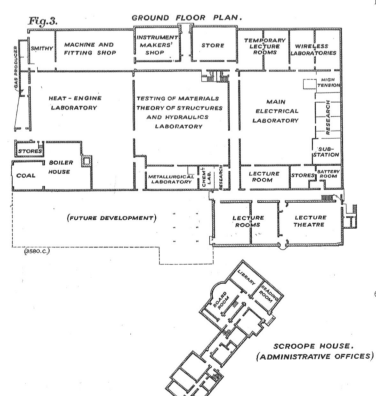

Fig.3.   GROUND FLOOR PLAN.

GAS PRODUCER

SMITHY | MACHINE AND FITTING SHOP | INSTRUMENT MAKERS' SHOP | STORE | TEMPORARY LECTURE ROOMS | WIRELESS LABORATORIES

HEAT - ENGINE LABORATORY | TESTING OF MATERIALS THEORY OF STRUCTURES AND HYDRAULICS LABORATORY | MAIN ELECTRICAL LABORATORY | HIGH TENSION | RESEARCH

STORES | BOILER HOUSE | | | SUB-STATION

COAL | | METALLURGICAL LABORATORY | CHEML. LAB. | RESEARCH | LECTURE ROOM | STORES | BATTERY ROOM

(FUTURE DEVELOPMENT) | LECTURE ROOMS | LECTURE THEATRE

(3580.C.)

BOARD ROOM | LIBRARY | READING ROOM

SCROOPE HOUSE.
(ADMINISTRATIVE OFFICES)

will automatically be picked up in a student's subsequent career are useful as stimulating interest, but apart from this they are of secondary importance.'

Inglis continued his personal research throughout his university career and beyond, but other members of the Department were so burdened by him with heavy teaching duties that they had no time for any other academic activity. Unlike Hopkinson, Inglis did not build up a group of PhD students around him, and kept teaching rather than research ability at the forefront when interviewing prospective staff members. He also detested administrative work and relied heavily upon two of his colleagues, J W Landon, Secretary of the Faculty Board, and A H Chapman, Secretary of the Department, to manage the Department. He believed research to be a private matter to be carried out by an individual for personal satisfaction only. Any attempt at building up a research group would require a great deal of administration, wasting time and energy. In this age of numerous research quality assessment exercises at universities, his attitude would be open to severe criticism!

Teaching quality was Inglis' mission throughout his career of 17 years under Ewing and Hopkinson and a further 24 years as Head of Department. He made few innovations and followed the path laid out by Hopkinson, firmly believing that mathematical analysis of engineering problems should always be checked by experimentation. His words on the purpose of education – 'the soul and spirit of education is that habit of mind which remains when a student has completely forgotten everything he has ever been taught' – were recorded in the speech he gave when he was inaugurated as President of the Institution of Civil Engineers in 1941.

ENGINEERING LEADERS: INGLIS' LEGACY

Some of the engineering students at Cambridge during Inglis' tenure gained great fame in later life for their achievements in engineering and many others made very valuable engineering contributions in the management of industry: Frank Whittle developed the jet engine; Morien Morgan became known as the 'Father of the Concorde'; the notable scientist G I Taylor worked in the Department for some time; Constance Elam (later Tipper) was the first female academic in the Department; James Goodier was appointed Professor of Applied Mechanics at Stanford University; Beryl Platt led a distinguished career in aeronautical engineering and then politics; and Jacques Heyman became Head of the Engineering Department in 1983.

The Engineering Department's Inglis Building is named in Inglis' honour, and to Inglis additionally goes the credit for founding the Cambridge University Engineers' Association, which encourages alumni to remain in touch with the Engineering Department. He persuaded Charles Parsons, who graduated from the University in 1877, to become the first President of the Association. Engineering's contribution to society was celebrated – Whittle, Morgan and Parsons were all knighted and Platt made a life peer.

Throughout his career, Inglis' teaching won the respect and admiration of all those who attended his lectures, during which he was said to have a 'magnetic attraction'. A former student wrote of him: 'He was the most inspired and inspiring of teachers that I have ever had, and this surely must be said by every man who had the supreme privilege of attending any of his lectures.' Another noted 'the joy with which he discovered – apparently for the first time – some vitally important factor in the course of a series of calculations involving two or three blackboards, which had to be experienced to be appreciated'. Inglis told a new intake of engineers: 'Your fathers, gentlemen, have sent you to Cambridge to be educated, not to become engineers. They think, however, that reading engineering is a very good way of becoming educated. In 10 years' time, however, 90 per cent of you will have become managers, whether of design, manufacturing, sales, research or even accounts departments in industry. The remaining 10 per cent of you will become successful lawyers, novelists, and things of that sort.'

But the golden years following the First World War were not to last, as for the second time in Inglis' lifetime the outbreak of war disrupted the University. Two decades of great progress came to an abrupt halt. Conscription came in again for undergraduates, who were called up for military duties in droves, although initially there were exemptions for engineers, scientists and medical students. Bursaries were created to maintain the supply of technical experts, but this dispensation ended as the war progressed. Many staff positions were left unfilled as more and more men departed for wartime duties. The workshop was used to produce munitions and for research on matters connected with

FRANK WHITTLE

Opposite: In 1964,
Morien Morgan in a BBC
documentary on Concorde
remarked: 'It's a lovely
shape – one feels that if
God wanted aircraft to fly
he would have meant them
to be this shape.'

Below: Frank Whittle
seated at his desk during
the Second World War.
The models on his desk are
the first British prototype
jet aircraft to fly and the
Meteor, the first jet to
enter service with the RAF.
Behind him is a model of
his jet engine.

Frank Whittle (1907–1996) precociously invented the turbojet engine when he was just 22 years old, describing it in a patent published in January 1930, and within his lifetime the invention transformed military aircraft and revolutionised civilian air travel. The first indication of Whittle's goal of devising a jet engine is found in a thesis he prepared at Cranwell as part of his officer training course. In it he concluded that aircraft would have to fly at very high altitudes in order to reach speeds of 500 mph and cover distances of thousands of miles, postulating that novel methods of power generation would have to be devised, such as rocket propulsion or gas turbines driving a propeller.

Whittle entered Cambridge as a mature student to read for the Mechanical Sciences Tripos in 1934, passed with first-class honours and, with encouragement from Melvill Jones, stayed an extra year as a postgraduate. Afterwards he formed a company, Power Jets Ltd, with two former colleagues from the RAF, and they raised venture funds for a contract with British Thomson-Houston (BT-H) to prepare design drawings. On 7th April 1941, the Gloster chief test pilot gave the W.1 engine its first run in the aircraft, which was airborne for a short period.

Whittle and his team designed the W.2 series of engines with better performance than the W.1. Rolls-Royce built and developed the W.2, installing it as the Welland engine in Meteor aircraft that were delivered to the RAF in May 1944 and went into action against German V-1 flying bombs. Apparently, they were able to keep up with the flying bombs and divert them away from their targets using the wingtips of the Meteor to flip over the flying bombs, disturbing the gyro-stabliser guidance system. Following further development, the W.2/700 was adopted by Rolls-Royce for the Derwent V engine, which claimed the world speed record of 606 mph in 1945.

Whittle's design of centrifugal flow compressors was gradually replaced with axial flow compressors and designs were improved using computers. Visiting Cambridge, Whittle was pleased to inspect the research into compressors and turbines in the Turbomachinery Laboratory, which was named the Whittle Laboratory in 1972. In his lifetime, he witnessed the first civilian jet airliner, the Comet, which made its maiden flight in 1949.

Below: Whittle's
student registration
card from Peterhouse.

wartime needs; as men left their positions in the workshop for war duties, a number of women were taught to use machinery and their work became an integral part of the war effort. Members of Queen Mary College, University of London were transferred to Cambridge and the engineering contingent amongst them used the Engineering Department for their teaching and research in the afternoons, when the Cambridge undergraduates had departed to work in their college rooms or for college supervisions.

Inglis retired officially in September 1940, having reached the age of 65, but continued as Head of Department for another three years. The Professorship of Mechanical Sciences was declared vacant and waited upon the appointment of the next incumbent.

CHAPTER FIVE

A Centre of Research Excellence

John Baker (1901–1985)

Professor of Mechanical Sciences (1943–1968);
Head of Engineering (1943–1968)

T he Second World War was still raging across Europe in 1943, but the prospect of victory for the Allied Powers was in sight and post-war reconstruction was under consideration across the nation. The moratorium on new appointments at the University of Cambridge was relaxed and a notice was inserted in the *Cambridge University Reporter* inviting applicants for the post of Professor of Mechanical Sciences and Head of the Department of Engineering. John Baker, Professor of Civil Engineering at Bristol University since 1933, had been seconded to the Ministry of Home Security in 1939 as Scientific Adviser to the Ministry's Design and Development Section, and it was at his Whitehall office that, out of the blue, a letter arrived from a Fellow of Clare College enclosing the notice in the *Reporter* and encouraging, 'almost commanding', Baker to apply.

A surprised and somewhat hesitant Baker overcame, as he later wrote, his 'modesty', and agreed to be considered for the position but questioned the meagre salary. Even at Bristol, where the Engineering Department under him comprised merely two lecturers and two technicians, one skilled and the other unskilled, his salary was greater. The reply firmly stated that the salary was non-negotiable and informed Baker that he would just have to manage in Cambridge with a lower standard of living! An inwardly delighted Baker applied, and shortly afterwards he received a telegram from the Vice-Chancellor of the University of Cambridge: 'You have been elected'. He overcame all his qualms, 'damned the financial consequences' and accepted the position. In August 1943 Baker was given permission to leave the Ministry of Home Security and take up his new academic position.

With the help of his younger colleagues in the Department, in particular R D Davies, Secretary of the Faculty Board, Baker formulated a clear vision for

By the end of his 25-year tenure, John Baker had transformed the Engineering Department. Research was expanded greatly and teaching modified to meet the demands of the post-war world. Alongside academic changes he oversaw a huge building project to provide much-needed new facilities for the Department.

Above: The Inglis Building developed in a series of projects during Baker's tenure. This photograph shows the finished building in 1966.

Left: Three-point bending test with electronic instrumentation and a camera mounted on the oscilloscope to capture the results.

Above: Suspension Bridge mural by Tony Bartl in the entrance of the Baker Building.

Below: Baker, wearing his signature bow tie, giving a lecture on the Morrison air raid shelter, which he designed using his plastic theory In 1958.

SAVING LIVES IN THE BLITZ: BAKER'S MORRISON SHELTER

One of Baker's duties as Scientific Adviser to the Ministry of Home Security was to examine the efficacy of air raid shelters, in particular the well-established Anderson shelter, which was designed for use in the garden. The study pointed out that most of the poorer communities in cities subjected to German bombing did not have gardens, and that even when they did have one, the shelter, partly buried in the ground, was susceptible to flooding in winter months. Baker was charged by the Home Secretary, Herbert Morrison, to devise a better shelter to protect families from aerial bombardment, without having to leave their homes. Immediately turning to his belief in plastic theory, Baker designed a shelter for indoor use that was about the size of a dining table and could shelter a family.

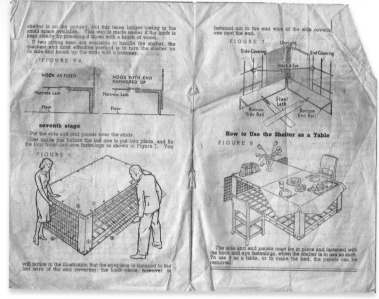

The shelter was delivered as a kit of several hundred parts that could be assembled by the recipient or by local Boy Scouts groups trained in the process. After assembly, the shelter comprised a box-like steel framework with a solid tabletop and wire mesh sides. When not needed as a shelter the sides could be removed and the steel structure would serve as a dining table or play area for the children!

Examination of bombed buildings had shown that there could not be any viable protection against a direct hit on a house but that bombs falling at some distance from a house either sucked or blew apart one or two walls, which caused the whole structure to collapse onto the ground floor, killing the inhabitants; the Morrison shelter was designed to withstand the consequence of this collapse. The energy of the debris falling onto the shelter was absorbed by the plastic deformation of the steel tabletop to a depth no greater than 12 inches, and Baker showed in his demonstrations that plastic deformation could absorb 100 times more energy than elastic deformation. More than one million shelters were distributed during the war years, and countless lives were saved. Baker received an OBE in recognition of this work.

In Cambridge, Baker liked to demonstrate the use of the shelter in his undergraduate lectures using scale models, and he gave a particularly dramatic demonstration by placing his gold watch under a steel bar on which a large weight was dropped. The bar deformed, the energy of the weight was absorbed in the plastic deformation and, thankfully, the precious watch survived.

Although Baker had invented and designed the shelter, it was expedient – or perhaps a political necessity during the war years – to name it after the Government Minister, Morrison. In 1950, when Baker met the Minister in Cambridge, there was an amusing exchange between them: Baker told the Minister that he was applying for an award for designing the shelter, to which the Minister replied in complete seriousness: 'But I thought I designed it myself.'

Above: Instructions on assembling a Morrison shelter.

Below: A Morrison shelter in use.

Below: At work in the small electrical laboratory, 1958.

Below right: A lecture highlight was Baker demonstrating the absorption of energy in plastic deformation by risking his gold watch placed under the deforming beam.

the framework of a post-war School of Engineering in Cambridge, preparing a 'statement of needs' designed to shape its post-war development. Distinctly aware that he was now working within the democratic structure of the University of Cambridge, he needed to ensure that his views were unequivocally shared by his Faculty Board. Working closely with Davies, he succeeded in obtaining the unqualified support of the Board, by Lent Term 1944, for a complete review of the Tripos structure, the introduction of postgraduate courses with an emphasis on the PhD degree (which Inglis had disliked), desk space for research students, a climate of research with modern experimental facilities, a doubling of the teaching staff in the Department and, as the first priority, a Chair in Electrical Engineering. In his strongly held views, wartime advances in all branches of engineering, but particularly in electronics, radar and communications, had made it imperative that Cambridge should have a visible presence in Electrical Engineering to complement the Aeronautical Engineering Professorship held by Melvill Jones and his own Chair in Mechanical Sciences.

There was no precedent in the University for such sweeping demands from a newly appointed Head of Department, but Baker had not only the determination and energy to achieve his aims but also extraordinary foresight. By the end of his 25-year reign he had transformed the Engineering Department, and he ranks indisputably as the greatest of the 14 Heads of the Engineering Department since its foundation in 1875.

The demands of the war had depleted the academic staff in the Engineering Department to just one professor and 24 lecturers by 1943, but by the time Baker retired, there were nine professors, five readers, 62 lecturers and demonstrators, and 35 other teaching and research staff supported by 223 assistant staff. The Department, hitherto known only for its teaching quality, was also recognised as a world-class centre of research.

In the course of his tenure, Baker brought some exceptionally able academics into the Department, and in making appointments during this period of unprecedented growth he insisted 'above all, that the candidate should be willing and able to do research'. Appointments Committees chaired by him were warned that they should only appoint candidates of the highest quality and, because the subject was engineering, they should expect candidates to have served their time in industry. Once the appointment was made, Baker supported his staff through their early years. His attitude towards colleges was somewhat ambivalent; while thoroughly enjoying his own Professorial Fellowship at Clare College, he warned newly appointed staff members against accepting college teaching positions before proving themselves in both teaching and research in his Department.

BAKER'S BUILDINGS

The visible manifestation of Baker's transformation of the Engineering Department was the rapid erection of buildings on the Scroope House Estate, on which he had found the original Inglis Building, a dilapidated Georgian house and temporary wooden huts. Shrewdly anticipating that national post-war reconstruction was imminent, Baker prepared his plans in advance of any formal announcement of the availability of funds for buildings and, having obtained the University's prior approval, he commissioned an architect, J M Easton, to prepare outline plans for the development of the whole of the site as well as more detailed plans for an extension to the Inglis Building.

Although the University authorities had informed departments that government funds could not be released until 1947, some money was made available in 1946, and as the Engineering Department's plans were ready, building work started immediately for workshops, laboratories and lecture theatres in an extension of the Inglis Building. Construction was completed in 1948 and the building was opened formally on 10th June 1949 by the Chancellor, Field Marshal Jan Smuts. A mural donated by Baker, 'A Short History of Engineering' by the artist Tony Bartl, was unveiled in the foyer of the Inglis Building. Baker's exceptional foresight and imagination had ensured that the first permanent post-war building authorised by the University was for the Engineering Department.

Emboldened by his success, Baker commissioned Easton to prepare plans for new buildings covering the whole of the Scroope House Estate, and he gained the approval of the University for the design of an E-shaped building with its spine running parallel to Trumpington Street and three wings extending towards

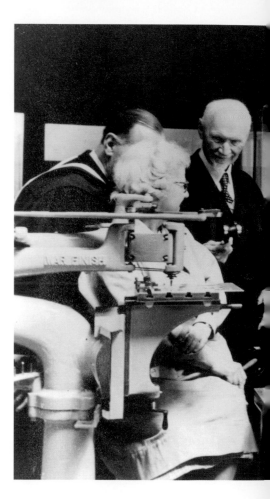

The official opening of the new workshops in the Inglis Building by the University's Chancellor, Jan Smuts, in 1949.

The mural, 'A Short History of Engineering', by Tony Bartl was given to the Department by Baker to mark the opening of the new workshops.

South-east view of the planned new workshop block.

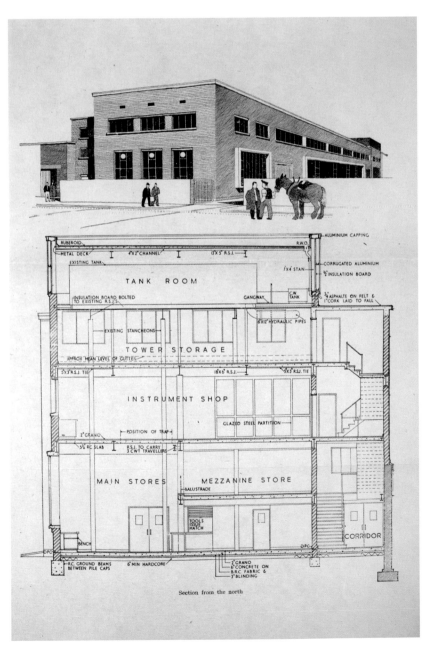

Section from the north

the Inglis Building. A one-storey laboratory was sandwiched between the South and Centre Wings and a paved area with an ornamental fountain broke up the density of buildings by forming a courtyard between the protruding North and Centre Wings. Although funds were available for this building in 1948, structural steel was strictly rationed and the amount available to the University was totally inadequate for its overall needs; inevitably, there were protracted arguments within the University as to which department should be allocated this meagre ration. The disputes raged on until the time limit for the utilisation of the steel was almost reached, but as Baker later wrote: 'At the psychological moment, Engineering came forward and said that there was no need to lose this steel

Painting by Terence Cuneo, presented to the Engineering Department by the Cambridge University Engineers' Association (CUEA) to commemorate the opening of the Baker Building in 1952 by HRH The Duke of Edinburgh.

A student in 1958 overseeing a wind tunnel experiment in the newly constructed Middle Wing of the Baker Building, temporary home of the aeronautical wind tunnels while the South Wing was under construction.

because we had our plans ready for our complete development so we could start right away, building just as much or as little as the steel would allow.'

Baker had anticipated events yet again, and the backbone of the E-shaped building designed by Easton was completed in 1952 and opened by His Royal Highness The Duke of Edinburgh in the presence of the Chancellor, Lord Tedder. The celebrations were 'a delightful occasion' attended by 250 guests and the building, a defining landmark of Baker's leadership, was named the Baker Building. Four storeys, a mezzanine floor and a basement provided the space needed for individual offices for all teaching staff, shared offices for the rising number of PhD research students, lecture rooms, teaching and research laboratories, committee meeting rooms and separate common rooms for students and staff. There was a Board Room for meetings of the Faculty Board in the style of the Board of Directors' Room in industry, a departmental library, storage space and administrative offices; the building became the envy of all University departments struggling to meet the post-war bulge in student numbers with inadequate buildings on cramped sites in the centre of Cambridge.

In 1952, there was further expansion of the Engineering Department when the newly appointed Professor of Aeronautical Engineering, Austyn Mair, together with the Professor of Applied Thermodynamics, William Hawthorne, asked for space for fluid mechanics and heat transfer research. They proposed that the South Wing of Easton's design should be erected to meet their needs and, within five years, Baker had another stroke of good fortune. A new government, in a period of national prosperity, decided that the time had come to provide funds through a new University Grants Commission (UGC) for the rapid development of technological education in selected universities, including Cambridge. Responding immediately, Baker put forward his prepared plans for the site, proposing that two wings of the Baker Building should be erected with a single-storey laboratory between them. The UGC provided funds to meet the entire cost of construction as well as of equipping the building for advanced technological research. In 1958, Aeronautics was accommodated on the top two floors of the South Wing, with Thermodynamics research on the first floor and the Fatigue and Control Engineering Laboratories in the Centre Wing.

Baker wrote: '[We had] the fun of designing the steel frames ourselves, trying out each time a more sophisticated method as it was developed by our structural research team.' The team was led by Michael Horne and Jacques

Above left: The Baker Building, 1952. It was one of the first buildings in the world to be designed using plastic theory. (Note the cladding in anticipation of further expansion.)

Left: Erection of the steel structure for the Centre Wing with Scroope House in the background, 1956. The house was demolished in 1962.

Heyman who, working with the plastic theory of design, saved 20 per cent in the weight of steel and reduced structural beam depths to 14 inches from the 20 inches specified by conventional elastic design theory.

In the 1960s, more space was needed for the electronics research laboratory, which had increased in size and moved into the dilapidated house on the site, to accommodate some of the research projects. Fortunately, Britain was still enjoying a golden age of prosperity, and money was easily available to add a North Wing to the Baker Building after demolishing Scroope House, thus completing Easton's original plan. Electrical research and teaching laboratories were accommodated in the upper floors of the North Wing in 1963. The North Wing was opened by Lord Nelson of Stafford, an industrialist who had graduated from the Department. The occasion was marked by the inauguration of the fountain, but when it was turned on soapsuds soon covered the entire courtyard, to Baker's visible fury! Mischievous undergraduates had surreptitiously emptied a packet of washing powder into the fountain. Baker soon recovered his good humour and was able to look back with pride and satisfaction that in 17 years he had managed to cover the whole three-acre site with buildings. His vision had finally been realised after two decades of endeavour, and he prophetically claimed that there was now sufficient space for the Engineering Department for the next 50 years of growth.

Towards the end of his tenure, Baker had yet another stroke of good fortune: a change in bye laws made it possible to build more floor area on the site than had been permitted by regulations in 1949. Parts of the Inglis Building were demolished to build Inglis A; when it was finished the ground floor was

North Wing and main courtyard of the Baker Building.

Below left: A steel structural experiment in the Structures Laboratory, 1958.

Below: Materials research, 1958.

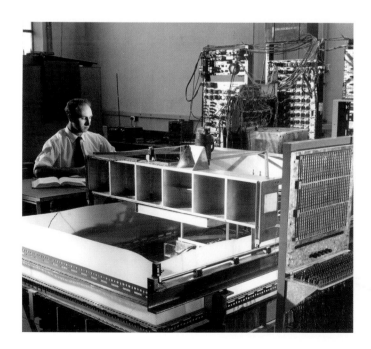

Christmas cards from the early 1950s, mainly featuring electronics laboratory research students. Produced by W Denys Cussins and Les Peters.

Cussins explains: 'The background for the 1952 card (**below**) was a local pawnshop. Henry was the senior lab steward and Uncle was Mr Oatley (later Professor Sir Charles Oatley). Driving the road roller was Firmin; whilst carrying off the young beauty was Dennis McMullan. Hanging from the corner of the shop is Short … Staggering away from his wrecked MG was Chris Grigson (in real life his MG was written off by a fire engine in Hull). Graham Clarke holds the salver on which rests the severed head of Ken Sander. Far right is myself looking as if I owned the Jaguar.'

occupied by the Structures Research Laboratory, the mezzanine floor by the Materials Teaching Laboratory, the first floor became a new Electrical Teaching Laboratory and the floor above was designated for electronics research.

Baker took enormous pleasure in embellishing his buildings with artwork, starting with Bartl's mural. He wrote that he took 'greater pride in this [the artwork] than in most other achievements of the twenty five years, a painting by Cuneo, two other murals, a portrait by Eric Kennington, an "abstract" of the concrete laboratory by Margaret Arthur and our recent brilliant aluminium construction by Kenneth Martin'. These symbols of his reign are still very much visible in the Department.

In Baker's 25 years, such was the rate of growth of scientific and technological education in the tradition-bound University of Cambridge that its members, with justification, feared that the dramatically shifting balance of resources between the arts and sciences would transform the University into an Institute of Technology in the pattern of MIT or Caltech in the USA. These fears were not allayed until the development of the arts campus on Sidgwick Avenue, now a spectacular panoply of modern architecture.

ACADEMIC ADVANCEMENTS

Among his demands, Baker had asked for a complete revision of both the Mechanical Sciences Tripos and the Ordinary Degree, while retaining the principle that all branches of engineering must be given the same weighting, although this would inevitably circumscribe the depth of teaching that could be provided in any one branch of the subject. A cautious approach was needed, because Baker knew full well that supervisors in colleges and private tutors earning a living in Cambridge by coaching undergraduates taught 'across the board', covering all branches of traditional engineering in their supervisions, and that they would resist changes that might compel them to acquire new or specialist knowledge.

The pressure for change was irresistible, however: during the war years, there had been unprecedented advances in technology, and new topics needed to be included and out-of-date lectures discarded if the quality of the Cambridge degree was to be maintained without burdening the students with work beyond their capacity. Knowing that this revision would be a controversial and contentious matter, Baker proposed that a Reorganisation Committee should be put in place and given the task of revising both the Tripos and the Ordinary Degree. Realising that the Committee would meet deeply entrenched positions among the teaching staff, he did not press for an early report.

Two years after its first meeting in 1943, the Reorganisation Committee proposed that the Tripos should be divided into two parts: Part I would be a 'general engineering' course without specialisation, to be taken by all undergraduates and constituting the academic qualification for the BA Honours degree, while Part II would allow for greater depth of study and specialisation into any one of the four well-established disciplines of engineering: civil, mechanical, aeronautical and electrical. The quality of entrants to the University had improved markedly since the war and so the Committee also proposed that the ablest should be permitted to take a fast course through Part I, completing it in just two years. These top students could then study Part II in their third year. The majority would follow the normal three-year course for Part I.

The standard of entry continued to rise and, 20 years on, Baker initiated 'a major reappraisal of the normal and fast courses leading to Part I of the Tripos' as increasing numbers of students wanted to take the fast course and specialise in their third year. Also some staff members wished to introduce new topics including computing courses and transistors in electronics instead of valves, and to transfer the optional exercises in structures into the compulsory syllabus.

In the 1965/66 academic year, computing was made compulsory for the first time, and the structures design project became a compulsory part of Part I, while the committees appointed to review the courses for the Mechanical Sciences Tripos continued to work towards a more radical revision of the Tripos. A Teaching Committee was established and, following intensive and sometimes acrimonious discussions, recommended the establishment of a new Engineering Tripos to take the place of the Mechanical Sciences Tripos.

John Horlock, who had studied and lectured in the Department before accepting a chair at Liverpool, returned as Professor of Engineering on 1st January 1967. He immediately took over responsibility for the Teaching Office and played a key part in finalising the recommendations of the Teaching Committee, in particular the recommendation to change the name of the Tripos. From 1967, all students entering the Department were accepted to sit for the Engineering Tripos, taking Part I in two years and one of the options in the third year: either Part II or Part II (General) or the Electrical Sciences Tripos. The 'Normal' course was eliminated through this reform, and eventually there was satisfaction on all sides, Part II (General) having quelled the fears of those who felt that engineers should not specialise before leaving Cambridge.

In 1968, there were violent student protests and widespread unrest in continental Europe and in the USA. In Cambridge the student protests were relatively mild and, although Senate House was illegally occupied for a few weeks and some damage done, there was no violence until the Garden House

Above: Flow visualisation, 1958.

Opposite: Gas inflow experiment, 1958.

OUT OF THE ORDINARY DEGREES

In 1963, the Reorganisation Committee proposed that the Engineering Studies course for the Ordinary Degree should be completely revised to make it more appropriate as a professional engineering qualification for 'non-mathematical' students whose minds were not suited to the discipline of the more analytical Tripos, but who would be fit to work effectively in the engineering industry. Some element of specialisation was allowed in the course and a management studies course was introduced. But the proposal smacked of a bygone age in Cambridge, when gentlemen could study for a degree course without too many intellectual demands and enter industrial management in possession of an Ordinary Degree. The course had to be swiftly abandoned, much to Baker's chagrin; he held surprisingly strong views on the merits of this course, and railed against the need for an 'Honours' degree for all.

The engineering industry was facing demanding technological challenges, and needed highly trained engineers; employers did not accept a qualification without an Honours degree and graduates with Ordinary Degrees could not find suitable positions. Also, by 1960, very few entrants to Cambridge were willing to take courses leading to an Ordinary Degree, and the Faculty Board formally recommended that the course should be abandoned. There were very few regrets when it was eventually terminated after almost 100 years, having evolved from the Special Examination in Mechanism and Applied Science, instituted in 1870.

Riot in 1970 (see the box entitled 'The Garden House Riot'on page 83). Baker wisely initiated a survey of students' attitudes to the Department and its performance on their behalf which gave the overall result that, although there was general satisfaction, the students felt that they were given insufficient information on their progress through the course of the year and that more staff–student exchanges were necessary.

Under Baker's regime, the Engineering Department not only grew in size but also modernised rapidly, with the teaching staff subdivided into groups roughly in line with the main areas of engineering: mechanical, aeronautical, electrical and civil. At the same time, a number of new lecturers were recruited, with engineering qualifications other than the Cambridge Mechanical Sciences Tripos. Baker marvelled at the ease with which he could recruit outstanding academics, but many of the newly appointed staff members in the Department and more widely in the University were not elected to fellowships by the tradition-bound colleges, leading to an increasing divide between those who were fellows of colleges and those who held only University positions.

Throughout his tenure, one of Baker's chief concerns was to find a way of reducing the hours taught by staff members, so as to enable them to devote more time to supervising the increasing number of students studying for the PhD degree without diminishing the quality of teaching for which the Department had been acclaimed during Inglis' tenure.

RESEARCH UNDER BAKER

Baker had arrived in Cambridge with a wealth of experience and with the determination to make his mark on the University. In Cambridge, he built new laboratories on a grand scale and firmly imposed new syllabuses on the entrenched and long-serving staff members; his greatest contribution, however, was his success in transforming the Engineering Department into a world-class centre of research. By the time he retired in 1968, there were more than 100 lecturers and six professors. Baker's ability to recruit outstanding academics

Left and above: The sculpture 'Construction in Aluminium' representing the formula for a helical screw propeller, by Kenneth Martin, commissioned by Baker. The sculpture was made in the Department's workshops and installed in the entrance to the Department in 1967. In 2016 it was granted Grade II status by Historic England.

John Earnshaw (with the pipe): 'I worked under Professor Dill Deck, a world expert on cathodes in high power microwave valves. This is my apparatus, built in the era before any solid state electronics was used. It was manufactured in the Department using valve technology. The photo was taken in 1965 and shows Dr Chris Maloney (left) and me.'

had a profound impact on engineering departments throughout the UK, and indeed the world. He insisted that the quality of undergraduate teaching had to be maintained at the highest level, and wrote that buildings on the site would only come alive if they were populated with leaders of original research of the highest quality.

For instance, Baker's interest in welding led to his appointment to the Welding Research Council, which supported some of his research work during the war years at Bristol University. Among his research staff was Richard Weck, who 'studied the residual stresses in welded joints and problems of metal fatigue'. Weck accompanied Baker on his transfer to Cambridge and immediately started to work in one of the small huts on the Scroope House site – there was then no other suitable space, as the Baker Building was yet to come. It was reported that 'noise from his experiments on resonant vibrations led to many complaints'.

The British Welding Research Association (BWRA), which was based in London, purchased a country house, Abington Hall, a few miles from Cambridge, and through the connection with Baker its research was closely linked to work in the Engineering Department. With Baker's continued involvement, BWRA built laboratories and administrative offices on the site, and Baker ensured that all the buildings were designed on the basis of plastic theory. Baker's principal collaborators on this work, in effect the design team, were Michael Horne, Jacques Heyman and Roger Johnson. In 1957, Weck left the Engineering Department to become Director of BWRA and widened the scope of its activities; its research went from strength to strength and it continues to thrive as TWI Limited.

CHANGES IN FACULTY

In 1943, one of Baker's priorities was to establish a permanent Chair in Electrical Engineering. Developments in radar, communications and electronics had placed the subject at the forefront of engineering, but in a climate of wartime restrictions the University did not have funds for a chair. Fortunately, the

BAKER'S LIFETIME RESEARCH: PLASTIC THEORY

In earlier years, engineers designed buildings, whether modern offices with steel frameworks or concrete housing blocks, to perform within the elastic limit – the limit of loading within which there is no permanent deformation. It was believed that this elastic limit also represented a safety limit, beyond which the structure might collapse. Baker realised that this understanding was flawed: modest plastic deformation does not immediately lead to catastrophic collapse, and the elastic requirement adds an unnecessary and costly feature to a design. Relaxing this requirement allows less material to be used in the construction, but to harness the full load-bearing potential of the materials without any compromise to safety required a full understanding of plastic deformation.

One of Baker's first actions on taking up his appointment in 1943 was to set up a well-funded Structures Research Laboratory dedicated to research on plastic theory. The team developed a design methodology based on Baker's early work in Bristol. As Jacques Heyman explains: 'The "elastic designer" attempts to calculate the actual state of the structure, so that he can assure himself that the building is safe, whereas the "plastic designer" makes a trivial inversion of the design statement: instead of requiring a building to stand up, he requires it not to fall down.'

Once this thinking is accepted, then the 'plastic designer' changes focus: instead of thinking about how a building would behave under its normal loading, they concentrate on how it could possibly collapse under an overload. The concept of collapse through permanent plastic deformation, and the energy absorbed in the process, is most elegantly displayed in Baker's work on the Morrison shelter, described earlier (see the box entitled 'Saving lives in the Blitz' on page 56). Baker also realised that building materials needed to have the ability to absorb construction imperfections or some play in the fitting of one section to another. Steel, the most widely used building material, and reinforced concrete, the modern designers' building material, both exhibit this ductility and experiments on loaded steel frameworks demonstrated that the plastic collapse load is almost always reproducible even when the fit between parts of the framework is grossly imperfect. This effect gave plastic designers even greater confidence in the long-term robustness of their efficient structures.

As the culmination of two decades of work, Baker published a two-volume account of plastic theory research and the design of the steel skeleton with Cambridge University Press, the second volume in co-authorship with Michael Horne and Jacques Heyman. The publication was the final seal on his life's research. As a consequence of practical demonstrations, the British Standard Specification was modified in 1948 to include plastic design. A few British structural engineers and architects began to use plastic theory to design their buildings, although civil engineers are conservative by nature and more conventional design methodology remained dominant for many years.

Above: Cartoon on the 'elastic limit' published in the CUES *Journal* 1921.

THE FIRST PROFESSOR OF ELECTRICAL ENGINEERING

Eric Moullin (1893–1963) read Part I Mathematics at Downing College before transferring to the Mechanical Sciences Tripos. After graduating, he worked in the Engineering Department for 10 years as an Assistant Director of Research based at King's College before moving to Oxford as a reader. At the outbreak of the Second World War, he was transferred to the Admiralty and later to the Metropolitan Vickers Electrical Company.

The foundation of a Chair in Electrical Engineering brought Moullin back to Cambridge. Described as 'a very original and creative research worker', he wrote a number of key scientific papers and books. As a lecturer, he was greatly admired and was noted for reinforcing his lectures with experimental demonstrations. He fostered the research environment desired by Baker in the Electrical Engineering Laboratory, but he did not believe in university research conducted with PhD students. His views are contained in the address he gave to the Institution of Electrical Engineers on his inauguration as President in 1949. Unfortunately he suffered from ill health for many years before his retirement in 1960 and he died three years later.

Below: Delegates at the Conference in Engineering Plasticity, 1968. A dinner was given in honour of Baker. It was deemed to have been 'an excellent conference with many great contributions from the world's top researchers'.

Institution of Electrical Engineers offered to pay the salary of the Professor for a period of five years, and Eric Moullin, who had worked previously in the Department and was a reader at the University of Oxford, was appointed as the first Professor of Electrical Engineering at Cambridge in April 1945. Baker launched an appeal in 1948 seeking a permanent endowment of the Chair of Electrical Engineering as the first priority and a Professorship of Applied Thermodynamics as his second priority. He wrote to his friends in industry and to former members of the Department: donations were received steadily and reached £100,000 by 1950; a further gift of £70,000 from the British Electrical and Allied Manufacturers Association made the total sufficient to endow a Chair in Electrical Engineering. The industrial company Imperial Chemical Industries donated funds for the Chair in Applied Thermodynamics, which supplemented the money received earlier for a lectureship from Mrs John Hopkinson, and William Hawthorne was appointed to the Hopkinson and ICI Chair in 1951.

In 1945, a Chair was endowed by the Shell Group, and it was understood initially that Chemical Engineering would be a part of the Engineering Department but, somewhat surprisingly, the subject was established as a small independent department. The first Head of Department was Terence Fox, a lecturer in the Engineering Department who was promoted to the Chair. He held a remarkable record as an undergraduate, not only winning the Rex Moir Prize for the best performance overall in the Mechanical Sciences Tripos but

Electrical Research Laboratory, c.1968.

Back row from left: R D Jackson, R L Ferrari, H Ahmed (the author of this book), B Bohan, J Cooke, J Brown, R Gill.

Front row from left: J Maddrell, P Mackenzie, A F Tubby, A H Beck and K F Sander.

also gaining prizes for the best performance in thermodynamics, aeronautical engineering and structures. Sadly, he suffered from poor health and gave up the headship of Chemical Engineering to return to the Engineering Department before dying prematurely at the age of 50. The Chemical Engineering Department recruited students mainly from the Natural Sciences with a few from the Engineering Department. The second Head of Department, Peter Danckwerts, remained in post for 16 years; his successor, John Davidson, served for 18 years and introduced a three-year course of Chemical Engineering. Chemical Engineering has grown steadily through its history and reached the highest ranks of international stature. In 2008, during the tenure of Lynn Gladden, the Department was amalgamated with the Institute of Biotechnology and changed its name to the Department of Chemical Engineering and Biotechnology.

For the first nine years of Baker's reign over the Department, Jones held the Professorship of Aeronautical Engineering, working in huts remote from the main Engineering Department (see the section entitled 'Aeronautical Engineering' on page 49). On Jones' retirement, Austyn Mair was elected to the Professorship and Aeronautical Engineering became an integral part of the Engineering Department. A Professorship in Mechanics was established in 1961, the first holder being Dan Johnson, who had previously worked in the Engineering Department as demonstrator and lecturer; he resigned his position of Professor of Mechanical Engineering at the University of Leeds to return, but stayed only for one year

Terence Fox (1912–1962) became the first Shell Professor of Chemical Engineering at Cambridge.

The wind tunnel in the Aeronautics Laboratory in the South Wing was opened by William Farren in 1963. From left to right: Austyn Mair, William Farren, John Baker and Melvill Jones.

before moving to a position in industry. His successor was Edward Parkes, who had graduated in the Mechanical Sciences Tripos in 1945 and worked at the Royal Aircraft Establishment. On 1st October 1965, two Professorships of Engineering were awarded, to Bill Beck and John Horlock. Baker had brought a wealth of intellectual talent into his Department and these committed academics, from professors to demonstrators, very quickly delivered research of high quality.

This was also the period in which the government, mindful of the technological advances in the USA and in Europe, decided the time had come to encourage the teaching of high technology in universities. When the Engineering Department requested a very substantial increase in the numbers of academic and assistant staff all requests were met, and the expansion of the Department was well underway by end of the 1950s. As more funds became available, there were numerous appointments to lectureships as well as junior academic positions, and also recruitment of supporting staff such as technical assistants and administrative officers. A Superintendent of the Workshops was appointed and by the end of the academic year 1964/65 there were five professors, six readers, more than 50 University lecturers, two University demonstrators and seven Assistant Directors of Research. There had also been a substantial increase in technical assistants and laboratory administrator posts, rising to 178. The Department reinforced its position as the largest in the University, driven by Baker's extraordinary energy and single-minded desire to create a centre of engineering research within a somewhat bemused University.

University regulations require professors to retire at the age of 67, but Baker was not ready for this vacuum in his life, only relinquishing his post with obvious reluctance on 30th September 1968. Nevertheless the time had come, and his retirement event was described in the Departmental report to the University: 'A soirée was held to mark the occasion, at which members of the staff and their ladies paid their respects to Sir John and Lady Baker.' The Cambridge University Engineers' Association held a separate function and showed their appreciation by presenting a small bronze piece by the celebrated sculptor Henry Moore and a painting of flowers by Anna Zinkeisen; the presentation was made by Lord Nelson, President of the Association. There was also a dance in the laboratory, at which Baker was given a photograph album recording the development of buildings and laboratories on the Scroope House Estate.

Sir John and Lady Baker receiving gifts at his retirement soirée in September 1968.

THE VANISHING LIBERTY SHIPS: CONSTANCE TIPPER IN THE ENGINEERING DEPARTMENT

A fleet of almost 3,000 seagoing freighters, described as 'ugly in appearance' but well suited to the purpose of carrying desperately needed supplies to Europe, was built in great haste during the Second World War in American and British shipyards. Named 'Liberty ships' after President Roosevelt quoted 'Give me liberty or give me death' at the launch of one of the ships, they sailed across the cold Atlantic waters to Britain and Russia under repeated attack from German U-boats. To the consternation of the shipbuilders, some ships mysteriously vanished without any sign of enemy attack. During investigations, it was observed that a small crack could suddenly develop in the ship's hull and then run almost instantaneously round the hull of the ship. In one instance, a ship actually split in two as soon as it was launched into the sea while still anchored within the harbour.

The Admiralty approached John Baker, recently appointed Head of the Engineering Department, to investigate the problem and determine the failure mechanism, informing him that, to speed up production, the ship's hulls had been built by welding instead of riveting the steel plates. At first, localised failure of welds was suspected to be the root cause of the cracking and Baker immediately started an elaborate research programme on welding stresses, but it soon became apparent that welding could not possibly be the primary source of the problem. Very sensibly, Baker had assigned the task of metallurgical investigations to Constance Tipper (née Elam), who was working at the Cavendish Laboratory with G I Taylor on the deformation of metal crystals.

With her pioneering and elegant research, Tipper showed that it was brittle fracture of the steel used to manufacture the ships rather than the welding that was to blame for the initiation of the failure mechanism. She demonstrated that embrittlement of the steel began when the temperature of the sea fell below a critical point. The ductile properties were instantaneously lost so that the brittle steel, behaving almost like cast iron, split at high stress points and the crack could then propagate rapidly. An outcome of the work was the well-known 'Tipper test', used as the standard test for determining the notch brittleness of steel.

But despite her evident talents in the field, her path was not an easy one. A telling account of the problems faced by women in engineering and science was given in a report on the Bakerian Lecture of 1923. The lecture, given annually at the Royal Society, has been used to report some of the most important research in the physical sciences, and Taylor and Elam were invited to speak on their groundbreaking work on crystal plasticity. An extract from an account by the Royal Society reads:

'The first dinner of the Royal Society club attended by a woman was in 1972, when Dorothy Hodgkin was invited as Bakerian Lecturer. Fifty years earlier, in 1923 the Bakerian Lecturers G I Taylor and C F Elam were invited to dine

Above: Constance Tipper (1894–1995) specialised in the investigation of mechanical properties of metals and their effects on the performance of engineering components and structures. During the Second World War, she worked on the causes of catastrophic fractures in Liberty ships.

Right: The *SS Schenechtady* suddenly cracked in half in January 1943 while moored in Oregon. Subsequent research concluded that the low-grade steel used in the tanker's construction was prone to brittleness in cold conditions.

Second World War poster from the United States advertising war bonds and promoting an opportunity to see a Liberty ship in person.

before it was discovered that the second lecturer was in fact Constance Elam. Rutherford suggested that she be sent a box of good chocolates, and Constance Elam replied graciously, declining the invitation and perhaps consuming the chocolates! The rules of the dining club were changed in 1923 to specify that if a Bakerian or Croonian Lecturer 'should happen to be a lady' she should be invited to dine. In 1974 it was resolved that women Fellows of the Royal Society could be proposed and admitted to the club.'

Constance Elam was more than a pioneering scientist!

After her marriage to George Howlett Tipper in 1928, Elam (now Tipper) settled in Cambridge, and Newnham College awarded her a Research Fellowship for one year (1930/31). In 1936 she was awarded a Leverhulme Fellowship, and Newnham College kept up a connection with her throughout her life, but during all this time she held no official post in the Department. She was, however, allowed to use the original kitchen in Scroope House as her office.

Tipper worked in the Engineering Department during the war years, when the staff was much depleted and the lecture load on her was very considerable, but sadly she did not enjoy lecturing, nor was she a capable lecturer. On one occasion, students who attended her lectures drove her out of the lecture theatre in tears. When the Baker Building was completed, one of the amenities for the staff was a common room, but women – and Constance Tipper was the only woman academic – were not allowed to use it. The appointment to Reader in Mechanical Engineering in 1949 was to give her much pleasure, however, as she had felt keenly the lack of any status in the Engineering Department, and she held the Readership until she retired in 1960.

Tipper's research is now considered to be, as Baker put it, 'one of the principal contributions to the development of metallurgical science', and there is no doubt that she was one of the most talented members of the Engineering Department – but one who did not receive enough credit in her working life, perhaps because she was a woman working in the male-dominated environment of engineering. After Tipper had retired and left Cambridge, Baker wrote that her contribution was 'outstanding work both scientifically and in its effect on the development of welded ships and she deserves the highest credit for it'. Liberty ships continued to sail across the Atlantic and made a vital contribution to the liberation of Europe from German occupation during the Second World War.

SEEING THE FUTURE: OATLEY AND THE SCANNING ELECTRON MICROSCOPE

Charles Oatley was the Acting Superintendent of the Royal Radar Establishment (RRE), poised for promotion to Superintendent, when he was approached by Baker to join the Engineering Department. Oatley turned down the opportunity to follow in the footsteps of John Cockcroft at the RRE and came to Cambridge, accepting a newly created engineering fellowship at Trinity College in 1945 and then a University Lectureship. Oatley was exactly the sort of recruit Baker was seeking, committed to research and with a background in microwave and radar technology, the key wartime developments in electrical engineering.

Oatley began to develop an experimental research programme based on his belief that 'a project for a PhD student must provide him with good training and, if he is doing experimental work, there is much to be said for choosing a problem which involves the construction or modification of some fairly complicated apparatus'. Initially, he faced opposition from Moullin, but an exchange of letters with copies to Baker brought resolution and Oatley was clear to enact his plans.

Arguably, the greatest achievement in the Engineering Department in Baker's era was the post-war development, by Oatley and his students, of the scanning electron microscope (SEM) invented in Germany in the 1930s by Manfred von Ardenne (which builds an image by detecting the emissions caused by scanning a tightly-focused beam of electrons across an object). The SEM project was launched in the 1950s by Dennis McMullan, Oatley's first research student, and the microscope is now ubiquitous, with thousands in use across the world. Described as 'the single most important scientific instrument of the post-war era', the SEM was initially ridiculed by microscopy experts, who denigrated it because of its comparatively low resolution – these critics failed to appreciate the value of SEM images, which contained a wealth of information obtained with little or no specimen preparation. All criticisms disappeared in the years that followed: voltage contrast proved invaluable for inspecting electronic 'chips'; dynamic observations in

Charles Oatley (1904–1996), pioneer of the scanning electron microscope.

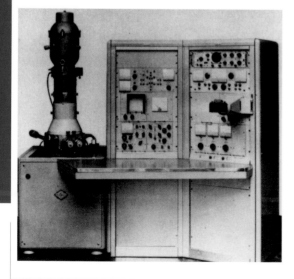

Far left: Scanning electron microscope SEM1 designed by Oatley and his first research student, Dennis McMullan.

Left: Ian M Ross graduated in first position in the Mechanical Sciences Tripos in 1952. Oatley and Moullin disagreed on whether Ross' future lay in industry or academia. Ross achieved success in both arenas: he stayed on at Cambridge to work on current fluctuation phenomena in semiconductors; and he later joined Bell Laboratories in the USA where he rose to the post of President (1979–1991).

environmental chambers gave dramatic insights into hitherto unexplained physical, chemical and biological processes; medical researchers imaged bacteria and even viruses using more recent high-resolution, scanning electron 'nanoscopes'. Oatley's vision and determination were recognised and rewarded.

Oatley's initial microscopes were difficult to operate and prone to breakdowns but nevertheless a number of research students used them very effectively for pioneering research. Ken Smith built the first SEM with magnetic lenses and took it to Canada in 1956 to study pulp and paper specimens. Tom Everhart worked on voltage contrast and paved the way for the exploration of new contrast mechanisms in the SEM, which were of particular importance for the examination of functioning microelectronic devices, and went on to become President of Caltech. Subsequently, a notable list of outstanding students worked with Oatley on scanning electron microscopy projects; the last of these students was Alec Broers, later Head of the Engineering Department and Vice-Chancellor of the University of Cambridge.

Oatley's initial attempts at commercialising the machine were unsuccessful, and a great deal of time was lost while it was believed that the worldwide market for SEMs would be six machines! In the hands of the Cambridge Instrument Company, a saleable product – the Stereoscan – was developed under the leadership of Gary Stewart, another of Oatley's outstanding students, and the first batch of instruments was sold very quickly in 1965. It was now apparent that the microscope provided a uniquely powerful method of inspection, and the market expanded rapidly. Cambridge Instrument's successor, ZEISS Microscopy, is still in Cambridge on the same premises producing SEMs to this day, having expanded significantly in 2013.

Modern scanning electronic microscope in use at the Institute for Manufacturing, studying the development of new materials and production processes.

CHAPTER SIX

Hawthorne to Heyman

William Hawthorne (1913–2011)

Hopkinson and ICI Professor of Applied Thermodynamics (1951–1980)

Head of Engineering (1968–1973)

On Baker's retirement, the terms and conditions for the appointment of the Department Head were revised; the practice of appointing to retiring age was abandoned and an initial period of appointment was fixed at five years but could be extended up to a maximum of five more years by mutual agreement between the incumbent, the Faculty Board and the University's General Board. These rules were introduced throughout the University and implemented whenever long-serving Heads of Department reached retirement age. In another significant change from past practice, the link between the Chair of Mechanical Sciences and the headship of the Department was abolished. By 1968, there were prominent professors in new academic disciplines and the entirely sensible decision was taken to abandon a restriction that would have denied them the opportunity of appointment to the headship.

A Search Committee was formed by the Faculty Board, with John Reddaway, Secretary of the Faculty Board, acting as Committee Secretary. It did not search for long nor very far from Cambridge and proposed to the Faculty Board that William Hawthorne, the most distinguished member of the Engineering Department at that time, was the most appropriate person to succeed Baker. Hawthorne, the Hopkinson and ICI Professor of Applied Thermodynamics, was well known for his research achievements, and highly regarded for his capacity for managing academic activities, directing major research laboratories in Cambridge and at MIT with conspicuous success. The choice was widely applauded both within the Department and in wider University circles until it was announced that Hawthorne would also be the next Master of Churchill College, Cambridge, succeeding John Cockcroft; he would thereby hold three major responsibilities simultaneously, and there were concerns that he would not be able to devote enough time to the management of the Department.

William Hawthorne's most outstanding work at Cambridge was in the understanding of loss mechanisms in turbomachinery. During his time as Head of Department, he established the Turbomachinery Laboratory with John Horlock.

Above: Austyn Mair at the controls of the wind tunnel he designed and installed in the South Wing of the Baker Building. The tunnel is still in use today.

Above right: Before becoming Head of Department in 1983, Jacques Heyman was responsible for the repair and restoration of the Great West Tower of Ely Cathedral in the early 1970s.

Right: Alec Broers operating his experimental apparatus while a PhD student in the Department.

Above: Materials Laboratory with J H Brunton, K J Pascoe and S Wright.

Left: Engineering Tripos Part 1 in 1974 with Mair as Head of Department.

Right: Department group photograph from 1967.

Back row from left to right: A F de O Falcão, R E Davis, D G Gregory Smith, B L Hirschman, A J T Horrath, B T Johnson, McH Nicholson, R Walmsley, D McLean, Maj J May, N Walker, P J Chappell.

Front row from left to right: H Merryweather, M J Ashdown, D Lindley, W R Hawthorne, H Marsh, J Heyman, R P Johnson, M C Yally, W E Coles, R D Greenwood, A K Mitka.

Photographs from the Undergraduate Design Course – the lowermost features Donald Green (centre).

Hawthorne pointed out that, historically, professors at the University of Cambridge were expected to run their own independent departments, but engineering professors were unusual in that they worked as a team in one department. Hawthorne understood how the University worked better than anyone, and so he devised a management structure to harness this *esprit de corps*, comprising two Deputy Heads, a Secretary of the Faculty Board and a Secretary of the Department.

APPLIED THERMODYNAMICS

Earlier, in 1951, when first elected Professor in the Department, Hawthorne made an immediate impact and started to build up a research group. His early research students included two outstanding men, John Horlock and Arthur Shercliff, who were both appointed to the staff of the Engineering Department after completing their PhD degrees. With their help, Hawthorne developed a new syllabus for undergraduate courses in thermodynamics, based on a strong analytical foundation and intellectual rigour derived from the lectures given at MIT. At the same time, he continued his pioneering research in fluid mechanics and turbomachinery, both in the Department and at MIT, with the support of outstanding teams of academic staff and very able research students. The results were of immense value to manufacturers of turbomachinery and jet engines on both sides of the Atlantic. Despite his appointment at the Engineering

SEA SERPENTS ON THE OUSE: THE DRACONE PROJECT

The Suez Crisis in 1956 created a worldwide shortage of oil, a problem for which Hawthorne proposed a novel solution: a flexible and collapsible barge for transporting oil around the Cape of Good Hope while the Suez Canal was blocked. The barge ended up as a sausage-shaped nylon tube which, when filled with a fluid, floated in seawater and could be towed by an oil tanker or other vessel to the point of use of the liquid. When the tube was floating in the sea, bulging pressure waves surged up and down the length of the tube, making it appear to be a 'huge snake swallowing a goat'; it was appropriately named the Dracone (from the Greek for a sea serpent). Initially, uncontrolled oscillations from side to side were encountered and the nylon fabric had to be strengthened, but research in the Department gradually helped to solve all operational problems. The tubes could be hundreds of feet long and tens of feet in diameter when full of cargo, but when empty they could be rolled up into small, relatively light packages for transport and reuse.

The re-opening of the Suez Canal and the construction of supertankers put an end to the need for Dracones to transport large quantities of oil. However, niche markets developed and a few are still in use, for transporting oil to locations where suitable docks for large tankers do not exist, as well as for transporting drinking water, petrol, kerosene and other fluids. Today, Dracones are also found to be useful when recovering waste oil from a spillage at sea, and they were used to good effect during the Falklands conflict in 1982. Some have been in operation for many years around the Greek islands for transporting fresh water and other liquids. In Cambridge, many generations of research students carried out research projects connected with Dracones, sometimes gathering together at jolly parties on the River Ouse at Ely, which was used for testing Dracones, known irreverently as 'Hawthorne's NOBs (nylon oil barges)'.

Right: A Dracone, carrying 26,000 gallons of oil, making an experimental trip on an inland waterway.

Below right: Hawthorne with Nigel Cubitt looking at a scale model of a Dracone in a wave tank, 1956.

Department, Hawthorne's association with MIT continued throughout his working life, first as Jerome Hunsacker Visiting Professor in 1955 and then, for much of his career, as Visiting Institute Professor, the highest rank of professor at MIT, working in the Institute's Gas Turbine Laboratory where he had a major role in guiding the work of a number of students, faculty and research staff. He always retained an office at MIT for his personal use.

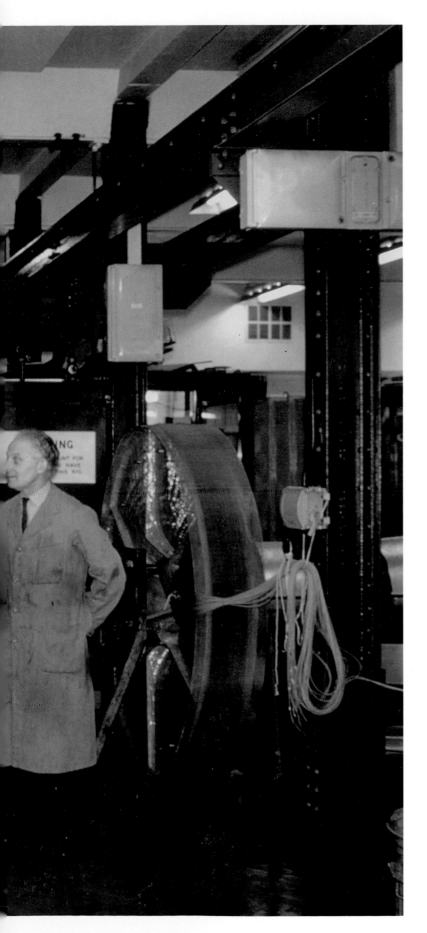

The Deverson rig was installed on the first floor of the South Wing before being transferred to the Whittle Laboratory. From left to right: J Furness, H Daneshyar, D S Whitehead, E Deverson, H Marsh, W R Hawthorne, J Horlock and A Carter.

A NEW ENGINEERING TRIPOS

Hawthorne had assumed charge of the largest department in the University, with almost 1,000 undergraduates, 160 research students, more than 100 academic staff and 200 technicians and administrative staff – a formidable responsibility. As he himself acknowledged, he was fortunate in having his former student and close colleague John Horlock back in the Department as Professor of Engineering and Deputy Head, and the two men promptly formed a close working alliance. Hawthorne said of Horlock: '[He] was a man in a hurry. He was never satisfied with the usual responsibilities. He liked solving administrative problems – and he'd got a lot of push.' Horlock needed all his 'push' as he continued in the role he had taken up in Baker's time of leading the revision and modernisation of the syllabus for the Tripos.

The structure and content of the Mechanical Sciences Tripos had been put in place when Baker arrived at the Engineering Department in the 1940s, and by the 1960s, as Horlock pointed out, there were both internal and external pressures for a radical revision. Internally within Cambridge, colleges were finding it difficult to advise candidates on the relatively complex structure of the Mechanical Sciences Tripos, which, uniquely in the University, offered the qualifications for a BA degree after just two years. Externally, industry was demanding 'increasing specialisation in the education of engineers entering research and development', but at the same time industry also asked for some engineers with a practical approach and expertise in design and manufacture, and Horlock noted that 'surprisingly few Cambridge engineers are using a slide rule 10 years after graduation'; most Cambridge graduates moved rapidly into positions of management in industry.

The reform, perhaps long overdue, was initiated by the Faculty Board under Baker's guidance in 1964 and discussions on committees and planning continued until 1967. The Revision Committees, nominated by the Faculty Board,

STUDENT VOICES RAISED AND HEARD: THE STAFF–STUDENT JOINT COMMITTEE

Student protests in continental Europe and the USA in the 1960s spilled over to the UK and specifically to Cambridge in 1969 when some of the Mill Lane lecture rooms and the Council Room of the Old Schools were occupied by protesters. At the Engineering Department, protesters directed volleys of paper darts at lecturers, ostentatiously displayed themselves reading newspapers during the lecture and drummed their feet to drown out the lecturer. Their main demand was a greater degree of student participation in governance of the Department's teaching and in response the Faculty Board set up a Staff–Student Joint Committee (SSJC), under the chairmanship of Horlock, comprising six students and six staff members, to consider all matters affecting undergraduate teaching. Students were invited to attend Faculty Board meetings for items of business concerning student affairs and were entitled to receive minutes of the Teaching Committee's deliberations. It was noted by Hawthorne that in the first few months of operation '[the SSJC] has already made a major contribution to teaching in the Department'.

Open Meetings were held in May 1970 and again in May 1971, both chaired with consummate skill by Hawthorne; attendance exceeded 250 and the lecture theatre was packed to capacity. Criticisms of teaching content, style and quality as well as the lack of amenities in the Department came not only from students, but also from some staff members. As a result, audio-visual aids were installed in lecture theatres, sales of second-hand textbooks were held in the Department and arrangements were made for more interaction between staff and students. From the point of view of some students and staff members, the best new feature was that a common room with coffee machines was made available to all! It was agreed that the Open Meeting would be an annual event and elections were held for student members of the SSJC with almost 50 per cent of the students voting in the elections. By 1972, the SSJC and the Teaching Committee had prepared questionnaires designed to improve the quality of teaching in the Department and these were generally thought to have been effective in improving the lectures. Changes were made to the assessment of laboratory work and the practical examination arrangements were cancelled. Lectures on the use of English, given by a member of the English Faculty, proved very popular and examiners noted that these efforts by the Department had improved 'the standard of the written word'.

In the early 1970s, after student pressure, the Common Room was opened to both staff and students and a coffee machine was installed.

were advised to introduce new topics into the lecture courses and to discard outdated material, sometimes against strong opposition from a cohort of long-established and traditionally minded lecturers. Programmed learning was to be introduced, the laboratory work was planned to be an integrated course in experimental engineering leading to project work in later years, and computing would feature strongly together with a significant design element in the exercises set in the drawing office.

Particularly strong disagreements had arisen in Baker's time when the long-established name the 'Mechanical Sciences Tripos' (MST) was replaced by the 'Engineering Tripos' (ET), with a majority of members of the Teaching

Scenes from the Garden House Riot, Cambridge, 1970.

Courtesy Cambridge News

THE GARDEN HOUSE RIOT

The Garden House Riot in February 1970 dramatically altered the nature of student protest in Cambridge. The event took place during 'Greek Week' (a celebration of Greek culture and historical contributions to humanity) in February 1970, when Greece was under the regime of the 'Colonels'. Demonstrators invaded the grounds of the Garden House Hotel, banged on windows to disrupt the speeches and invaded parts of the building, causing considerable damage until a cohort of policemen with dogs put an end to the 'riot', but not before some policemen and a University Proctor had been injured. The local newspaper called the disturbance a riot and some of the national press printed rather lurid stories. Students were arrested and University Proctors were forced to divulge the names of University members at the protest. Some students were sent to prison for periods ranging from 9 to 18 months, while overseas students were recommended for deportation; many commentators felt that University members had been very harshly treated. After serving their sentences, some students returned to complete their degrees. Not only was this a life-changing event for those who were imprisoned, it also subdued protests by students in the UK for several decades.

Courtesy Cambridge News

HRH The Prince of Wales visiting the Engineering Department while he was an undergraduate at Trinity in 1969.

Committee successfully arguing that a change in name was necessary because Cambridge graduates were often labelled by industry as 'mechanical engineers' not qualified to work in either the civil or the electrical engineering industries. The idea of a graduate in 'mechanical sciences', who was nonetheless capable of contributing effectively in any branch of engineering, was neither understood nor accepted by industry at a time of greater specialisation, with entrants into industry from other universities holding clearly demarcated degrees in electrical, mechanical, civil or aeronautical engineering. A degree in 'engineering' was much easier to explain to prospective employers as a qualification in general engineering. The change was of particular benefit to Cambridge graduates seeking employment outside the UK during the emerging globalisation phenomenon, which saw the 'export' of British engineering expertise through multinational companies employing Cambridge graduates for work on engineering projects across the world, from the Middle East to Australia. When all was decided, the accolade in *The Times* read that Cambridge had created 'an Engineering Tripos to please the critics'.

Machine tools in the workshops.

John Coales (seated third from left) with members of his Control Group alongside the hybrid computer.

The formal start of Hawthorne's headship, 1st October 1968, coincided with the introduction of the new Engineering Tripos, and Horlock stressed that two conditions that define Cambridge engineering had been met: the new Tripos did not separate the Engineering Department into civil, mechanical and electrical departments, and the course was firmly based on the fundamentals of engineering science, ranging across all engineering disciplines. Horlock wrote that 'Cambridge graduates make their mark as men who can tackle any job that arises in the engineering industry', and he quoted as examples the contributions to jet engines, atomic power, the plastic theory of structural design, flexible oil barges and the scanning electron microscope.

Horlock made a considerable impact in each of his two tenures at the Engineering Department. Working closely with Hawthorne in the 1950s he led the revision of courses in thermodynamics before moving away from Cambridge when he was appointed Harrison Professor of Mechanical Engineering at Liverpool University at the age of 30. Hawthorne was determined to entice him back to Cambridge and in 1967, in the penultimate year of Baker's term of office, Horlock was appointed to a Professorship of Engineering and to the position of Deputy Head of Department for Teaching.

Horlock led the reform and reorganisation of the Mechanical Sciences Tripos to the new Engineering Tripos, countering the opposition from the 'Old Guard' who were against almost any change. Later in life, he wrote that this period of his life in Cambridge was not altogether a happy one, partly because of tensions in his relationship with the academic staff and also because of the inadequate experimental resources for turbomachinery research in Cambridge compared with those he had left behind at Liverpool University. The first cohort of undergraduates reading the Engineering Tripos completed their first year in 1969 using the new syllabus that required a relatively high level of mathematical expertise. There was some concern that the majority of undergraduates had found the mathematical content too demanding, and a Working Party on Teaching Methods was appointed to review the level of mathematical skills expected from entrants to the Department. Detailed descriptions of mathematical topics considered important for an engineering degree,

COMPUTING

The importance of computing in engineering was recognised at all levels early on, and the IBM 1130 computer, heavily used for research work, was upgraded and linked to the IBM 370/165 University computer based in the Mathematical Laboratory (renamed the Computer Laboratory in 1969), greatly enhancing the computing power available to Engineering Department users. The Department's electromechanical calculating devices were replaced with a programmable Hewlett-Packard 9810 installed in the student drawing office for general use, and it is hard to comprehend today the excitement throughout the whole Department at the availability of this now seemingly trivial computing service – the size of a small suitcase, its computing capabilities were equal to little more than a basic four-function calculator today. Nevertheless, students acquired computing skills with great enthusiasm, and it quickly became obvious that some formal instruction would be necessary within the taught courses in the Tripos.

Below: The IBM 1130 console.

Bottom: Researcher using the Control Group hybrid computer.

together with examples of typical problems, were sent to Directors of Studies in colleges for circulation to schools and also to the cohort of undergraduates preparing to enter Cambridge in 1969. Working together, Hawthorne and Horlock guided through a very substantial modernisation of engineering teaching in Cambridge in the early years of Hawthorne's headship.

ADVANCES IN RESEARCH AND CHANGES IN FACULTY

The active research topics in the Department listed in the annual report by Hawthorne in 1968/69 were applied thermodynamics and gas dynamics; control engineering; electrical engineering; fluid mechanics; management studies; materials; mechanics; soil mechanics; and structures. In 1972/73, acoustics was added with the arrival in the Department of John (Shôn) Ffowcs Williams, an authority on noise reduction in aircraft. Hawthorne noted in his report that a grant of £470,000 had been awarded to the Control Engineering Group under Professor John Coales, and a grant of £50,000 from the Wolfson Foundation had been awarded to the Electron Optical Group within the Electrical Division, but that there was a shortage of space for accommodating these new initiatives. An interesting part of Hawthorne's report concerned the survey of graduates from the Department over the past 20 years: 3 per cent had migrated to the USA and 10 per cent to other parts of

Administrative team from the 1960s.

the world, but almost all were working for British firms operating abroad, showing that if there was any 'brain drain', as suggested in the alarmist press of the day, the phenomenon did not apply to Cambridge engineers.

Research activity in the Department was growing at an ever-increasing rate, and is perhaps best assessed by noting that members of the Department published 750 papers in academic journals or conference proceedings during the five years of Hawthorne's headship. The Department's growing reputation as a world-class centre of research brought numerous distinguished academics from across the world to give research lectures or to carry out collaborative research programmes with their colleagues in the Department.

As an example of the rapid growth of research, the work of the electronics research laboratory was described by the author in the Cambridge University Engineering Society's *Journal* for 1969. Seven semi-independent research groups headed by 15 academic staff covered such diverse topics as electron optical instruments; microwaves; plasma; the Gunn Effect; applications of superconductors; physical electronics; thermionic and photo emission; tunnel emission; electrical control systems; high-power mercury arc rectifiers and inverters; instrumentation and communications; and computer-aided design (CAD). Other divisions were equally active and outstanding research was carried out in all the 10 sub-divisions listed earlier.

CAD research was carried out in collaboration with the University Mathematical Laboratory (later the Cambridge Computer Laboratory), headed by the pioneering computer scientist Maurice Wilkes.

CAD research was carried out in collaboration with the University Mathematical Laboratory, headed by Professor Maurice Wilkes, with the work divided between software development in the Mathematical Laboratory, and design analysis followed by production of 3D mechanical components in the Engineering Department. A computer system, the Elliott 905, was installed in the Department and linked to the CAD Centre, which had been established in Cambridge by the government. In 1970/71, the six years of collaborative activity funded by the Science Research Council (SRC) for joint research by the Engineering Department and the Mathematical Laboratory were brought to an end, and the Department made a separate application to the SRC for Engineering Design with Computer Supervision.

Appointments to chairs and promotions to senior positions are valuable indicators of the growth of research activity in an expanding department. In November 1970, there were nine professors in the Department, six readers, 61 lecturers, 14 assistant directors of research, seven senior assistants in research and five demonstrators, and all of these, with very few exceptions, were active in their personal research and the supervision of PhD students. Baker's transformation of the Department into an Engineering Research Laboratory was now manifestly obvious.

Professor Ken Roscoe created a very significant research activity on soil mechanics and, although he was tragically killed in a road accident on 10th April 1970, his successors built strongly on this foundation (see Chapter 9). Peter McGregor Ross was appointed Professor of Engineering on 1st July 1970 to fill the vacancy created by Roscoe's untimely death. Professor Oatley retired in 1971, having worked in the Department for 26 years, during which he made a name for himself and brought great credit to the Department with his work on the scanning electron microscope. In 1971, Peter Brandon was appointed to the Professorship of Electrical Engineering vacated by Oatley.

John Ffowcs Williams was elected to the Rank Professorship of Engineering, transferred a major

Above: Construction of the Schofield Centre centrifuge in 1973. The 10-metre centrifuge was designed by Philip Turner, Senior Design Engineer in the Department from 1959 to 1984.

Left: A Xerox Sigma 6 was installed in the basement below the fountain in 1971, giving local access to mainframe computing to the undergraduates.

research activity from Imperial College to the Department, and began to build up new research projects. Mike Ashby was elected to a Professorship of Engineering (1966) from his professorship at Harvard and continued his work on materials and the important subject of materials selection (see Chapter 9). The death was recorded in 1972 of Harry Rhoden, Reader in Engineering and long-serving member of the Department from 1939, preceding Baker. He was the last link to the conditions in the Department during the war years.

A number of readerships were created, which helped to retain gifted research engineers in Cambridge, preventing them from going to chairs in other universities. Jacques Heyman was promoted to a Readership in Engineering in recognition of his many contributions to the plastic theory of design over a period of two decades, working first with Baker and then independently. Malcolm Head was appointed Reader in Aerodynamics in recognition of his work on air flow and boundary layers. Ken Johnson was appointed Reader in Engineering – a somewhat belated recognition of his research in the field of contact mechanics. Denis Whitehead was appointed Reader in Engineering in recognition of his work on supersonic flow along cascades of blades. Frank Fallside was promoted to a Readership in Electrical Engineering in recognition of his work on control systems and the interaction of humans with computers (see Chapter 9).

In 1973, at the end of his five years as Head of Department, Hawthorne did not seek a further term in office; although he was used to working remarkably long hours and working more efficiently than most of his peers, he was stretched to the limit with three simultaneous responsibilities, serving as both Head of Department and Master of Churchill College, and guiding the work of a very significant research laboratory at MIT, which frequently required his presence in Cambridge, Massachusetts. Consulting assignments for industry and memberships of government committees and committees of the Royal Society kept him more than fully occupied. To many members of the Engineering Department, he always appeared to be about to take a flight to Boston or a train to London.

There was a telling episode during Hawthorne's retirement event, when the lecturer Henry Cohen, speaking about Hawthorne's life, produced three bowls placed face down. He lifted the first one with nothing hidden underneath, exclaiming: 'Is Will at the Engineering Department? The answer is NO.' On lifting the second, the question changed to: 'Is Will at Churchill College? The answer is still NO.' With the third bowl, the question became: 'Is

Will at MIT? And the answer is still NO', he exclaimed, 'because Will is seated in front of us, enjoying his retirement party.' Hawthorne *did* enjoy the occasion, taking the Cohen performance with great good humour. Hawthorne remained Master of Churchill College until the retirement age of 70 and continued with his research, both in Cambridge at the Whittle Laboratory and at MIT, for many years after his retirement.

Horlock expected to succeed Hawthorne, but Mair proved to be the more popular candidate among the academic staff, and in 1974 Horlock abruptly left Cambridge to take the position of Vice-Chancellor of the newly founded University of Salford. Cambridge's loss was the nation's gain, and Horlock rendered great service to education in the UK, first at Salford, where he was instrumental in overseeing the University's progress, and then as Vice-Chancellor of the Open University from 1981. There he introduced Master's degrees, strengthened the courses in science and engineering and created an Open Business School, achievements that were recognised with a knighthood in 1996.

Above: Frank Whittle being given a tour of the SRC Turbomachinery Laboratory at its official opening in 1973. Located on the West Cambridge site, it housed low-speed and rotating cascade tunnels moved from the Heat Laboratory as well as a 2 MW compressor.

Frank Whittle and John Horlock at the opening of the SRC. Horlock oversaw the development of the laboratory.

William Austyn Mair (1917–2008)

Francis Mond Professor of Aeronautical Engineering
(1952–1984); Head of Engineering (1973–1983)

During his headship, Austyn Mair oversaw many improvements to laboratory facilities, the appointment to a number of new engineering professorships and the introduction of a new four-year course for the Production Engineering Tripos (later known as the Manufacturing Engineering Tripos).

On 1st October 1973, the tall, slim, upright figure of Austyn Mair was seen cycling into the Engineering Department at 9am precisely, reporting for work in the style he had adopted ever since his appointment as Francis Mond Professor in 1952 – but on this date he was arriving to take up responsibility for the Department for the next 10 years. His appointment had been uncontroversial; his colleagues and peers had selected him by popular acclaim, mindful of his distinction as an aeronautical engineer, his long service in the Department (more than two decades) and, most importantly of all, his reputation for fairness and objectivity. As an undergraduate in the Engineering Department, Mair had come under the influence of Melvill Jones, Francis Mond Professor of Aeronautical Engineering, and the experience persuaded him to make his own career in aeronautics – an inspired choice, considering that less than two decades later he would occupy the chair of his mentor.

Following the retirement of Jones in 1952, Mair, to his immense gratification, was able to return to Cambridge as the second holder of the Francis Mond Professorship of Aeronautical Engineering; he was then 35 years old and joined Baker, Hawthorne and Moullin as one of just four professors in the Engineering

A 'Wingsail' triplane in the test section of the large low-speed wind tunnel in the South Wing of the Baker Building, 1982. David Ireson is sitting in the test model and the photograph was taken by his final-year project partner, Keith Thomas. Thomas explains: 'The project was for John Walker Wingsails, to characterise the aerodynamic performance of triplane arrangements of rigid slotted sails as proof of concept for large cargo ships.'

When Mair was appointed Professor of Aeronautical Engineering in 1952, Cambridge's Aeronautics Laboratory comprised three small, low-speed wind tunnels in a wooden hut. Mair designed and built a supersonic wind tunnel and a larger, low-speed wind tunnel, which are still in use today in the South Wing of the Baker Building (**right**).

Department. Mair was aware that his predecessor, Jones, had worked throughout his career in a hut leased from the University Air Squadron, located on a site remote from the main site of the Department on Trumpington Street, and Mair inherited this 'laboratory', which included three low-speed wind tunnels, entirely unsuitable for the research he was planning. It must have been a nightmare for him, the prospect of yet again having to be content with a research laboratory separated by some distance from his teaching duties as it had been at Manchester University. Mair had been convinced throughout his career that experimental work should be strongly linked to theoretical research and teaching, and was adamant on this occasion that his research laboratory should be located in the same premises as his group's theoretical research and teaching activities.

Working closely with Hawthorne, who had been appointed to a chair just one year earlier, he argued for the construction of the South Wing of Baker Building to house the wind tunnels that he needed for his research. Working expeditiously, he produced a report for the Faculty Board, 'Notes on a Proposed New Fluid Mechanics Laboratory for Mechanical and Aeronautical Engineering', which described the installation of new wind tunnels and also proposed a close working relationship with fluid mechanics and heat transfer research under Hawthorne. To his immense disappointment, he was told that there was no money to implement his proposal, and the Faculty Board merely endorsed the report, but Mair's decision to prepare the report while fearing that there was little chance of executing it was to prove inspired.

As described in Chapter 5, there was a change in government policy in 1957, when the government offered increased grants to universities, enabling the Engineering Department to implement Mair's proposals, which were ready and waiting. Clearly, the timing had been perfect. The South Wing of Easton's E-shaped Baker Building was built for aeronautics, and not for the first time the complaint 'science first and the rest nowhere' was heard in the University – but this time 'science' was replaced by 'technology'. The South Wing was completed in 1958, six years after Mair's appointment, but it was worth waiting for: state-of-the-art research facilities, including a high-speed, supersonic wind tunnel and a low-speed wind tunnel, both of which were in operation by 1960. Over the next two decades, Mair built up a team of academics and students dedicated to research in aeronautics, and in 1973 he was appointed Head of Department.

Mair's personal research in this period included a detailed study of vertical take-off and landing (VTOL) and short take-off and landing (STOL) aircraft, focusing particularly on the possibility of such aircraft reducing journey times between cities. Mair pointed out that much of the time for a journey by air was actually spent travelling to and from an office or home in the city to the airport, and speculated that the journey time between capital cities, for example the centre of London and the centre of Paris, could be reduced to one hour provided suitable landing areas were made available in central sites.

The hovercraft, invented by Christopher Cockerell, a former member of the Department, was also being prominently investigated throughout the 1960s. Mair delivered a paper on the subject, 'Physical Principles of Hovercraft', to the Royal Aeronautical Society in 1964 and, working with Hovercraft Development Ltd, he examined the problems caused by wind forces on hovercraft and high-speed trains. He continued to work in this general area of research and delivered the Lanchester Memorial Lecture at the Royal Aeronautical Society in 1966 entitled 'STOL – Some Possibilities and Limitations'. Mair's interest and expertise in the subject led to his appointment to the Chairmanship of the Powered Lift Committee of the Aeronautical Research Council, but it became apparent in the 1970s that such aircraft would only be developed for military purposes, and indeed the Hawker Siddeley Harrier aircraft which flew in the 1960s, amongst other military career aircraft, has been developed since then. Civilian use was not pursued, due to improvements in road transport and the development of high-speed trains.

Christopher Cockerell (1910–1999), inventor of the hovercraft, watches the *Princess Margaret* speed up the Thames 1979. (In 1985 she was blown on to a breakwater at Dover and four passengers were killed.) In 2000 the *Princess Margaret* and her sister craft, *Princess Anne*, were taken out of service. They are the only surviving examples of SRN4 hovercraft. In 2016 the Hovercraft Museum in Hampshire launched a high-profile campaign to save one of the craft from being scrapped, to avoid 'a truly wonderful and unique piece of British maritime heritage being lost for ever'.

CAMPAIGNING FOR MORE WOMEN ENGINEERS

Throughout the decade of Mair's leadership of the Department, there was concern that very few women were reading engineering, although large numbers were being admitted to former all-male colleges of the University. In an effort to redress the imbalance, the Department collaborated with the Engineering Industry Training Board in running a four-day course for schoolgirls, known as INSIGHT, as part of the national campaign to encourage girls to read engineering at British universities. The success of this and the WISE initiative, which both focused on making the engineering profession attractive to women, is demonstrated today, when more than 25 per cent of undergraduates in the Engineering Department are women.

In 1966 this modified Auster T.7, designated a Marshall M4, crashed during a test flight – the Department was greatly affected by the deaths of the pilot and the PhD student on board.

One of the most difficult and painful periods in Mair's life arose following the crash of the Department's research aircraft, the modified Auster T7, which flew from Marshall's airport for flight tests piloted by test pilot Brian Wass. The plane had been modified by Marshall Aerospace by adding perforations in the wings and flaps to draw in air for a gas turbine, and it had been used for some years for in-flight experiments designed to explore boundary layer control by suction. It crashed for reasons that could not be ascertained with any certainty, on 8th March 1966, killing both the pilot and the young PhD student, Ramaswamiah Krishnamurthy, who was carrying out investigations in the course of the flight. Although Mair was not directly supervising the student, who was working under the guidance of Malcolm Head, Lecturer in Aeronautics, he felt personally responsible for Krishnamurthy's loss, and shared keenly the sorrow of the grieving widow who returned to India; in-flight research was thenceforth terminated.

By the time of Mair's appointment to Head of Department, much of the unrest among the undergraduates experienced in Hawthorne's time had ended, and the vociferous demands of the cohorts admitted in the late 1960s had been replaced by a generation that was relatively indifferent to serving on academic or governance committees. (It was reported by members of the Staff–Student Joint Committee that it was proving very difficult to find students willing to serve on it.) Mair's tenure was a calm period on the Scroope House Estate in another sense: there was little or no new building activity, the completion of the Baker Building having filled almost all the space available on the three-acre plot. While assuming responsibility for the Department, Mair was conscious that, with the economic recession biting throughout the country, the years of exceptional funding for universities from the government had come to an end – perhaps for decades, in the view of some commentators.

EDUCATIONAL DEVELOPMENTS

Mair put forward a proposal to the Faculty Board for a thorough review of the Engineering Tripos, taught for the first time in the academic year 1968/69, pointing out that two cohorts of undergraduates had now used the new three-year syllabus and improvements were necessary. Revised syllabuses and laboratory teaching were gradually introduced but there was a moratorium on this activity when the Department was made aware that the government was dissatisfied with the low status of the engineering profession in the eyes of the nation compared with their counterparts in other countries. The government was also unhappy with the relatively short duration of the science and engineering courses in British universities compared with the four- or five-year-long courses taught in continental Europe. They believed that British engineers were ill-trained to meet the challenges that would arise in the 'high-tech' age that was coming to Europe from the USA.

CONNECTING ACADEMIA WITH INDUSTRY

In 1971, under Hawthorne, the Wolfson Cambridge Industrial Unit was founded with a capital grant of £110,000 from the Wolfson Foundation. Donald Welbourn was appointed the first Director (responsible to the Head of the Engineering Department) and given the remit to 'assist people both in the University and in industry to understand better how they can help one another, and by so doing to promote the welfare of the University as a place of education, learning and research, while at the same time increasing the wealth of the country'. The unit took off during Mair's term as Head as it attracted further funding from industrial sources and charitable foundations. For instance, it made significant contributions in the field of computer-aided design by developing numerically controlled machines for the manufacture of complex components. A programme, DUCT, developed in the unit, was exploited with great success commercially, and helped to treble the capital of the unit in its first 10 years. Some notable commercial companies adopted the programme, including British Telecom and the German car manufacturers Volkswagen and Daimler-Benz. Following the retirement of Welbourn in 1983, the Wolfson Cambridge Industrial Unit was reformed under Stephen Bragg, the second Director of the Unit, and expanded to cover all fields of investigation in the University.

The Callaghan government commissioned an enquiry in 1977 under the chairmanship of Monty Finniston. It reported its findings in January 1980, recommending that the duration of courses in engineering and science subjects should be four years, that graduates should be awarded a Master's degree rather than a Bachelor's degree, and that this qualification should be entirely distinct from the traditional Cambridge MA degree. The Faculty Board endorsed the Finniston Report and the Engineering Department decided to follow its recommendations by extending the Engineering Tripos from three to four years; planning started immediately to broaden the syllabus and to include management and policy-making in a syllabus dominated hitherto by engineering science. It was immediately obvious that it would take several years of difficult planning and preparation before the four-year Tripos could be implemented – indeed, four-year Triposes would not be adopted across the Department until 1992.

Above: Schematic overview of the DUCT System Mark 32.

Above left: Early use of DUCT to produce aerofoil sections on the new CNC (computer numeric control) machines in the Department workshop.

Another government initiative in 1977 offered Teaching Fellowships to selected universities in the UK. Cambridge was awarded a Teaching Fellowship in Electroproduction, funded by the Electricity Council, and Ricky Metaxas was appointed to the Fellowship, with responsibility for teaching courses on the generation and utilisation of electrical energy and for creating a close liaison between UK electrical industries and the Engineering Department. The appointment was a success, and in 1979 the Electricity Council extended the Fellowship by another five years. This Fellowship was in addition to an earlier appointment supported by the Post Office, and an additional Teaching Fellowship was created with the support of British Gas in 1981. Fellowships provided welcome assistance to the normal complement of University lecturers burdened with heavy lecturing and laboratory demonstrating duties.

One-year Master's in Philosophy (MPhil) degree courses were also initiated in Cambridge in the 1970s, in order to serve as intermediate qualifications between the BA and PhD degrees. These were distinctly different from the Cambridge MA, which was exclusively awarded to Cambridge Bachelors of Art without further study, six years after the first term of residence providing the Bachelor's degree had been held for at least two years. These new MPhils varied in design: some were taught entirely by lectures and awarded on the basis of written examinations and a dissertation; others were research-based and awarded on the basis of a dissertation and oral examination. The new courses instituted in the Department in the 1970s were an MPhil in Soil Mechanics, an MPhil in Control Engineering and Operational Research, and the MPhil in Industrial Systems, Manufacture and Management that had started life in 1966 as the Advanced Course in Production Methods and Management (ACPMM).

In recognition of the need to provide strong leadership, Mair appointed Donald Green to the position of Deputy Head of Department in 1980, with responsibility for teaching. Green had graduated from Cambridge in Baker's time, returning to the Department after a distinguished military career, and was known to be meticulous in administrative matters; he was also well known and trusted by the academic staff of the Department. Mair had chosen well: Green was vastly experienced both in Departmental issues and in the day-to-day business, not only of the Department but also of the University. He had attended meetings of the University's General Board and, when necessary, deputised for the Head of Department on University and Departmental Committees during the term of Hawthorne's headship, when he was serving in the post of Secretary to the Faculty Board. Mair and Green, two like-minded men, made a formidable team in managing the Department through some difficult periods in the early 1980s, and in driving through the reforms of teaching necessary for the implementation of the new Tripos.

Presentation by Mair to Alexander Barker (on the left) to mark 50 years of service in the workshops, 1973.

The ACPMM European Tour, 1979: outside the European Commission building in Brussels.

The ACPMM European Tour, 1979: Zaanse Schans in the Netherlands. From left to right: John Walton, Mike Williamson, Steve Dow, Martin Monahan and Steve Calvert.

TECHNOLOGY AND MANUFACTURING IN EDUCATION

Rapid advances in microelectronics technology in the 1970s enabled dramatic increases in the density of devices in integrated circuits and led to the invention of microprocessors, which, with appropriate control software, found a multitude of applications in industry. In the late seventies, the Department decided to introduce undergraduate courses on the use of microprocessors, with practical classes on microprocessors in the Electrical and in the Control Engineering teaching laboratories available only to first-year undergraduates, although these experiments were so popular that many second- and third-year students used the equipment in their own time. The practical courses were extended in the next year, when 16-bit microprocessors became available and optional lectures on microprocessors were offered to first- and second-year students. In his report in 1981, Mair wrote that the teaching of 'microprocessors was extended in several ways – a new experiment on the development of single purpose microprocessors was introduced for the second year "electrical option" and experiments using a powerful 16-bit machine were introduced in the third year'.

Undergraduate projects using microprocessors for a wide range of applications, some devised by students and others by staff, proved very popular, and a 16-bit machine was installed for third-year projects in the early 1980s. Eventually the lecture courses on microprocessors were made compulsory for all first- and second-year students. Undergraduates proved very receptive to teaching by new methods, and a programmed learning text on Laplace Transforms, prepared by the newly appointed Professor, Arthur Shercliff, was received enthusiastically.

Computer-aided design (CAD) was also introduced into the undergraduate engineering course, using six graphic terminals installed in the student drawing office alongside the conventional drawing boards, together with software from the DUCT design-and-manufacture package and the BUILD software packages developed by the CAD group. To the immense satisfaction of all members of the Department, these were made available for unsupervised use by undergraduates, Production Engineering Tripos (PET) students, research students and academic staff, and not only proved immensely popular but also produced generations of Cambridge graduates with the skill and expertise to make British industry more competitive in the age of computer-controlled systems and robotics.

DRIVING CHANGE IN THE UK INDUSTRY: THE PRODUCTION ENGINEERING TRIPOS

In the mid-1970s the Engineering Department responded to an initiative by the University Grants Commission (UGC) asking universities to work more closely with industry in preparing undergraduates for future employment in engineering companies. In response, a new four-year Tripos, to be known as the Production Engineering Tripos (PET), was proposed as an alternative to the Engineering Tripos.

Part I of the PET was to be the same as the two-year Engineering Tripos Part I, but in the third year PET undergraduates would be offered a series of entirely separate lectures, such as production technology, design for manufacturing production, organisation and controlled production systems, human behaviour, personnel management and industrial relations, and financial aspects of management – topics considered appropriate introductions to an industrial company's needs. In the fourth year of the Tripos, PET students would spend the first half of the year in the Engineering Department attending focused lecture courses on the branch of manufacturing most relevant to their future and the second half in an industrial company of their choosing. They would undertake a major project of eight weeks' duration and several short projects, all connected with industry, and finally make a number of short industrial visits, perhaps 20 in all.

It was proposed that Part I of the Tripos would be classed in the normal manner, but that Part II would be an unclassified honours examination, and that distinctions would be awarded for especially meritorious work over the whole of the course.

Planning for the new Tripos continued in the academic year 1978/79, under Mair's guidance, and the first intake of students was admitted in October 1979.

RESEARCH UNDER MAIR

The foundations of Department-wide research in independent research groups, each headed by a strongly motivated academic staff member, were laid down in the time of Baker, and during Mair's tenure research activity in all groups continued to flourish. There was a 10-year period of sustained growth, with more than 2,000 papers published and almost 350 PhD degrees awarded, including some to such notable individuals as Ann Dowling, later Dame Ann Dowling and Head of the Engineering Department, and Robert Mair, son of Austyn Mair, who in 2016 gained the distinction of being the first serving member of the Department to be ennobled.

In 1974 the Management Studies Group joined with the Control Engineering Division to form a new Control and Management Systems Division, which also included the CAD group. The whole of their activity was located on the University's Old Press site on Mill Lane where the Control Group had earlier built its home when shortages of space began to be felt in the Department.

Above: The ACPMM minibus, 1978.

Top: PET European Tour, 1982. From left to right: Simon Fenley, Sarah Baker (née Barrett), Christine Tacon, Phil Morgan, Phil Sorrell, Toby Benzecry and Chris Croft.

Three new staff members were recruited to the Engineering Department and an Advisory Panel comprising 13 senior industrialists was appointed to oversee the PET. In the first year, 13 students were admitted – 12 men and one woman – and 11 of these continued on to take Part II; the admission for the next year was 21 students and two more lecturers were appointed to teach the course. In 1982/83, the first students graduated by completing Part II, and a distinction was awarded to one student.

The popularity of the new Tripos was immediately established, with 31 students opting for the PET in the academic year beginning in October 1981 and, notably, one of the students was awarded a National Design Prize. Responding to the success of the course and the quality of the Cambridge students, industry began to support the Cambridge initiative by seconding Industrial Fellows to give welcome part-time support to the overstretched teaching staff in the Engineering Department. Companies such as Ford, Plessey, GEC, British Aerospace and IBM provided funds for the Industrial Fellows. The Department was particularly gratified when students from the 1981 cohort set up a high-tech company, Rhombus Ltd, after graduating.

Mair led the planning and institution of the PET in the Department with consummate skill, and in his last year as Head of Department he was able to report that the Tripos was beginning to establish a reputation for the quality of its training, had received a special mention in the government's report entitled *Advanced Manufacturing Technology* and had been singled out for praise by the President of the Institution of Mechanical Engineering when he was speaking at the launch of the Campaign for Increased Manufacturing Effectiveness.

By 1978, it was clear that more research space was needed for the Microelectronics Group, which had expanded rapidly because of the support from its research partners in industry – GEC and BT – and through large grants from the European Union initiative, known as the ESPRIT programme.

With Mair's support and the encouragement of Oatley, the Microelectronics Research Laboratory, headed by Haroon Ahmed, moved to new premises in the Cambridge Science Park owned by Trinity College. The Engineering Department's Microelectronics Laboratory was opened in 1982 by His Royal Highness The Duke of Edinburgh, Chancellor of the University. Later, it expanded into larger premises on the Cambridge Science Park, but was eventually transferred to a purpose-built laboratory as part of the Physics Department. The Whittle Laboratory and the Schofield Laboratory for Soil Mechanics also expanded during Mair's tenure, as described in detail in Chapter 10.

Operators' area with line printer and control desk for the Xerox mainframe computer in the basement below the fountain on the Scroope House site.

CHANGES IN FACULTY

Economic recession began to be felt across the nation in the mid-1970s, and government money ceased to meet inflationary increases in the University, which ran seriously short of operational funds. The time of plenty enjoyed by Baker and to some extent by Hawthorne came to an abrupt end just as Mair took charge of the Department, and he had to report to the Faculty Board that the University's commitment to reduce expenditure had resulted in the Department losing six posts – one professorship and five lectureships. The posts would have to be held in abeyance after they had been vacated through normal retirements at the end of the academic year.

As if this was not bad enough, Mair was told by the University that five posts, including two professorships which would fall vacant in 1982/83, were also in jeopardy (although, gratifyingly, the two professorships were restored in the next year). Faced with this difficult situation, Mair appealed to the teaching staff members, asking them to redouble their efforts to ensure that the Cambridge undergraduate's experience of an engineering degree did not diminish in quality. A number of academics who had taken early retirement to help the University balance its books in its difficult financial position came to the rescue

Above and opposite: The Lady Baker Memorial Sculpture by Peter Lyon was unveiled in 1981 by Viscount Caldecote. It was welded in the Department and engraved by David Kindersley.

as volunteers, some working almost full-time in the Department in support of Mair during the period of austerity. Mair also reported that assistant staff numbers would be reduced by 7 per cent over the next two years with no prospect of restoration to former assistant staffing levels in the future.

But throughout these straitened times, there were also improvements and advances. The professorial appointment of David Newland brought the new topic of design research into the Engineering Department, and it was expanded gradually into a major activity in both research and teaching. Another notable expansion of research followed the 1972 appointment of the first Rank Professor of Engineering, John (Shôn) Ffowcs Williams, who brought with him his research on anti-sound and on reduction of jet engine noise, specifically from the Concorde aircraft. This research, in collaboration with Rolls-Royce, had been started by Ffowcs Williams at Imperial College. In Cambridge, Ffowcs Williams went on to co-found a company, TopExpress, to exploit his research commercially, and he was elected Master of Emmanuel College, Cambridge in 1996.

Following the retirement of John Coales in 1974, Alistair MacFarlane was appointed to the Chair of Control Engineering, who established the Control and Management Services Division and, in 1983, the Information Engineering Division. His research was particularly concerned with multivariable frequency response methods, which became a prominent theme in the Department for several years. William Johnson, who was appointed to a Chair of Mechanics in 1975, established a research activity in experimental and theoretical solid mechanics, metal-forming techniques and for plasticity applied to forging. He took early retirement and left the Department in 1982.

In his last year as Head of Department, Mair was still reporting that the University's continued 'commitment to reduce expenditure had resulted in the suppression of five lectureships', but at the same time he was able to convey some

The Lady Baker Memorial Sculpture today.

In 1981, Mair recorded the honours awarded to members of the Department: a Royal Medal of the Royal Society was awarded to William Hawthorne, and the Sir Harold Hartley Medal of the Institution of Management and Control was awarded to Alistair MacFarlane, who was also elected to the Fellowship of Engineering (later the Royal Academy of Engineering). Ken Johnson and William Johnson were both elected to Fellowships of the Royal Society.

good news: the Department had been allocated four 'new blood' positions in strategic target areas of engineering under a government scheme administered on a nationwide basis by the Science and Engineering Research Council (SERC). It was Mair's unfortunate duty to manage the Department at a time of stringent cost-cutting compared with the plentiful supply of money in the first three decades after the end of the Second World War, and it took a great deal of his patient but authoritative management to take the Department through this period of crisis while managing to retain the full confidence and support of the staff at all levels, so much so that he was asked to stay on for a second term of five years in 1978.

Having reached the 10-year limit of his appointment, Mair retired from the headship in 1983, and from the Francis Mond Chair a year later. His period in office was marked by very significant advances in engineering education and, in his discreet but highly professional manner, he had led the Department with great success through a difficult period of economic recession, thereby fulfilling all the expectations of those who had supported his appointment in 1973.

John Arthur Shercliff (1927–1983)

Hopkinson and ICI Professor of Applied Thermodynamics (1980–1983); Head of Engineering (1983)

Arthur Shercliff, Hopkinson and ICI Professor of Applied Thermodynamics, was clearly the most academically distinguished candidate for the headship on the retirement of Mair, and he was duly appointed eighth Head of Department in October 1983, but unfortunately the opportunity of introducing his ideas into the Cambridge engineering degree course was curtailed. Shercliff had been diagnosed with an aggressive form of cancer in September 1982 and following treatment and recuperation he was appointed Head of the Department in October 1983, but sadly he died within three months of taking up office. There was just time for him to elect two professors – John Carroll to Professor of Engineering and Alec Broers to the Chair of Electrical Engineering – before he died on 6th December 1983. After his death, the Arthur Shercliff Memorial Trust was established at Warwick and Cambridge universities to promote technical visits abroad by students of engineering, recognising Shercliff's year at Harvard University to which he had attached great value.

Above: John Arthur Shercliff photographed in 1983. His son, Hugh, joined the teaching staff of the Engineering Department in 1994, and is a Senior Lecturer in the Materials, Mechanics and Design Division.

Right: Mair, Mr Wallace (Superintendent of the Workshops) and Shercliff (on the right) at the Head of Department handover party in October 1983.

Jacques Heyman (b. 1925)

Professor of Engineering (1971–1992);
Head of Engineering (1983–1992)

Jacques Heyman applied the plastic theory to the analysis of masonry buildings. This application of modern techniques into older masonry buildings made him the world's leading expert in cathedral and church engineering.

The unexpected death of Shercliff created an urgent need for a new Head of Department, and staff members, without exception, were delighted to learn that Jacques Heyman, their long-standing colleague, had agreed to step into the breach. Having come to Cambridge as an undergraduate in the 1940s, when Inglis was lecturing, and stayed on to rise to a Professorship of Engineering, Heyman knew the history and functioning of the Department. In his early years, he worked closely with Baker on plastic theory applied to steel structures, but later in his career he worked independently on the application of the theory to masonry and, well before his appointment to the headship, he had established an international reputation as a distinguished civil engineer. Members of the Department also held him in high esteem for the objectivity and judgement he demonstrated while serving on Departmental committees and on the Faculty Board of the Engineering Department. Indeed, until Shercliff was appointed to the chair vacated by Hawthorne, Heyman had been the favoured successor to Mair, but he made way for Shercliff, remarking with characteristic modesty that Shercliff was 'of a different order of brightness'.

On his appointment, Heyman's stated aim was 'to make the place as an efficient machine. I tried to be invisible – letting people get on with the research they wanted to do, without filling in forms'. Immediately after his appointment, he created an efficient administrative structure, based in offices close to the Head of Department's office on the mezzanine floor of the Engineering Department. Senior members of the administrative staff were promoted to University Officer level and he demanded from them a high order of skill and commitment to ensure that academic staff could pursue research and teaching without spending their valuable time on administration.

By the time Heyman took responsibility for the Department, there were 881 undergraduates, 191 research students working for PhD degrees, 16 students reading for the Production Engineering Tripos, 22 postgraduates studying on the Advanced Course in Production Methods and Management (ACPMM) and 54 postdoctoral research assistants distributed among research groups in the Department. In other words, like his predecessors from the 1920s onwards, Heyman had to contend with managing the largest department in the University. Just before his appointment, there had been a drop in staff numbers because of the imposition of compulsory retrenchment by the University, but 'new blood' and 'information technology' schemes, funded by the Research Councils, had added enough new staff members to show a marginal overall increase.

DISCOVERING WHY ANCIENT CATHEDRALS STAND UP

In 1966, Heyman emerged from Baker's shadow when he proposed that plastic theory could be applied to any kind of structure that exhibits ductile behaviour and published his article 'The Stone Skeleton', which explained how plastic theory may be adapted and applied to the field of masonry architecture – a milestone in the modern theory for masonry structures. In subsequent publications, the theory has been applied to many structures found in cathedrals and churches, such as vaults, domes, towers and spires. In a paper for the general reader, 'Why Ancient Cathedrals Stand Up – The Structural Design of Masonry', he explained simply but elegantly that the 'stress' requirements needed for the design of steel or reinforced concrete are transformed into 'geometrical requirements for the design of masonry' and used the stability of the masonry arch to demonstrate his ideas.

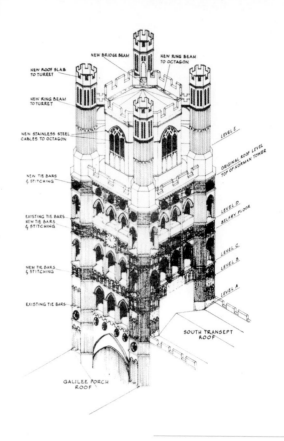

Above: Heyman's diagram of the work on the Great West Tower of Ely Cathedral for the Dean and Chapter in 1973.

Left: Cover of a leaflet outlining the work at Ely.

Below left: Heyman as a young man.

Two professors had joined the Department, nine new lecturers had been appointed, and there had been a considerable increase in unestablished posts, namely industrial, college and SERC Fellows. Fellowships funded by the Electricity Council and British Gas had been continued and provided valuable support in teaching and research. In the year after Heyman's appointment, a British Telecom Fellowship and new Industrial Fellowships sponsored by Barclays Bank, Jaguar and Unilever were awarded to the Department, providing further support in teaching and research. Heyman noted also that undergraduate numbers had increased by 20 per cent and postgraduate numbers by 30 per cent over the past 10 years. Postgraduate research staff had expanded even more rapidly as a consequence of large SERC Alvey research grants in Computer Speech and Information Engineering. He added that continual expansion of research was putting enormous pressure on the space available in the Department – a refrain that was to be heard over and over again for decades.

In his first year as Head of Department, Heyman was involved in the planning of three Tripos courses: the reformed Engineering Tripos, the Electrical and Information Sciences Tripos and the Management Studies Tripos. There

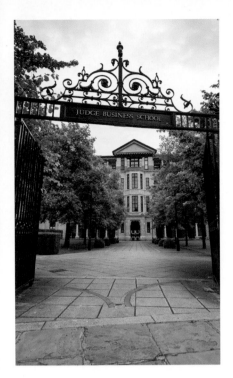

THE JUDGE INSTITUTE: A NEW START-UP FOR MANAGEMENT STUDIES

A generous endowment from Paul Judge, a University alumnus who studied Natural Sciences and Management Studies, enabled the University to establish the Judge Institute of Management Studies, 'a major development which will have a wide and influential effect'. Six new staff members were recruited, the MPhil in Management Studies admitted 19 students and the MBA at the Judge Institute attracted 20 students. Further benefactions enabled the University to develop the old Addenbrooke's Hospital site into magnificent accommodation for the University's activity in Management Studies, and the Engineering Department's teaching and research were planned to run alongside the work of the new staff recruited for the Judge Institute. Heyman wrote: 'Relations between the Judge Institute and the Engineering Department are intended to be particularly close, and the staff of the Management Studies Group in the Engineering Department and of the Judge Institute will be run as a single unit.'

was a common first-year course for all three degrees, but in the second year five core courses became compulsory and two further courses were selected from a list of four available courses in each Tripos. There was more emphasis on information engineering and electronics than hitherto, but the overall workload on the undergraduates was reduced. It was planned to use computing equipment for teaching purposes based on the results from the computer-based teaching project, supported by gifts from IBM, Acorn computers and ICL.

The Mathematics Group carried out a radical revision of the mathematics taught to first-year engineering undergraduates, with greater emphasis placed on numerical methods and formal procedures of analysis. A booklet describing pre-university mathematics was sent to all students accepted for the new Tripos. Heyman noted that, from the 1992 academic year, the entry of undergraduates would move from a three-year to a four-year programme, with the first two years concentrating on a broad education and the subsequent two years on a deeper understanding of a specialised discipline. Students would graduate with BA, MEng at the end of their four years in the Department.

Above left: Judge Business School, formerly known as the Judge Institute of Management Studies, is the business school of the University of Cambridge.

Below: The refurbished Engineering Library in 1992.

RESEARCH UNDER HEYMAN

Historically, there had been unprecedented growth of research activity in the Department in the decades following the Second World War as a result of a sequence of government funding for science and technology initiatives in universities. A period of exceptional growth in research activity was reported in the mid-1980s by Heyman, when he wrote in his report to the University:

'[The Department is] engaged in research of the greatest importance to industry and to the fundamental science of technology. There is no real shortage of money for this research – the University, the Research Councils, and, above all industry, are able and willing to fund work covering the whole range of engineering activity. However, the Department is now desperately short of space – space for the teaching of an ever-increasing number of students, and space for carrying out important research, from the solution of individual practical problems to the advancement of fundamental theory.'

He added that there were a number of active and important research projects in almost every field of engineering that could have been funded by external bodies but which had to be put on hold or turned down because of the shortage of space. His plea for more space is still being heard three decades later!

In his annual report for 1988/89, Heyman wrote that a four-year Manufacturing Engineering Course had been established in the Department, and there had been considerable expansion in electrical and information engineering teaching and research, with several new government-funded posts and newly appointed lecturing staff. Close contacts with industry, sponsorships contracts and generous donations of equipment had enabled the Department to continue to produce graduates of the highest quality.

Heyman identified six research areas that had grown organically within the Department: turbomachinery; soil mechanics and the geotechnical centrifuge; control engineering; microcircuit technology; optoelectronics; and computer speech and language processing. He concluded: 'The Department has never been stronger than it is now.'

The annual budget was reported to be £10 million, of which £3 million had come in the form of grants and research contracts. The available 25,000 m² of space was fully occupied, and temporary huts had to be built for an overspill into the grounds of the Department. Pressure on space had arisen from the need for an integrated circuit clean laboratory and the developments in speech, language and vision as applied to artificial intelligence. Work on robots, machine vision and the development of intelligent systems for automatic control of manufacturing processes was being carried out in cramped conditions, and the whole Department desperately needed more room. A second golden age of increasing research activities was being tarnished by the shortage of space, and there did not appear to be a solution to the problem.

Staff and students in 1983, featuring the following faculty members seated on the front row: Ann Dowling in the red jacket in the centre with Derek Smith to the left and, to the right, Jacques Heyman, then Austyn Mair, Dudley Robinson, Peter Brandon and Ken Livesley. Picking out a few students from the scene: Richard Prager, the University's current Head of the School of Technology, is standing in the third row behind Brandon; Adrian Dickens, on the front row wearing the jacket with a broad yellow stripe, founded the company Adder one year after this photograph, which continues to thrive under his leadership; and Mark Plumbley, seen laughing on the third row third from the left, is now the Professor of Signal Processing at the University of Surrey.

Heyman did propose a partial solution, which included refurbishing existing buildings and replacing single-storey laboratories with multi-storey space. Ageing apparatus should be disposed of and new equipment at the very forefront of technical innovation installed. Heyman wrote, in the year before he retired: 'These are fundamental needs if the Department is to keep its place as a leader in the advancement of engineering science and if it is to continue to produce outstandingly qualified graduates essential for the management of industry.'

Heyman's nine years as Head of Department was certainly a period of sustained growth in research, despite being hampered by the shortage of space: the number of research students rose by more than 80 per cent to 350, the number of postdoctoral research workers increased by 50 per cent to 74, and the number of papers published per year climbed from 247 to 456. Sixty PhD degrees were awarded in Heyman's final year as Head of Department, making a total of almost 450 during his term of office.

A BRIEF INTERREGNUM

The Cambridge Engineering Department nearly lost Heyman at the age of 38 when he accepted the newly appointed Chair at Oxford University. The appointment had promised great things to come, including space and equipment for research in new buildings. However, within a few weeks of his arrival it had become clear that funds were not available, and there were no plans in place to acquire any. Heyman resigned and created the extraordinary precedent of having held a chair for the shortest duration on record!

On returning to Cambridge, he found that his lectureship had been filled by Donald Green, and the University was not minded to create a new lectureship for him. Undeterred, he explained his predicament to Stephen Harris, the intrepid Secretary of the Faculty Board, who marched up to the Old Schools and persuaded the University secretariat to tear up his letter of resignation, which was still going through formal processing. There was now no evidence of a resignation, which meant that Heyman was able to retain his lectureship and another position had to be created for Green.

Below: Two photographs of the Mechanics Laboratory c.1985, with Hugh Hunt featured on the left in the front row in the lower picture.

CHANGES IN FACULTY

In 1984, Alec Broers was appointed to the Professorship of Electrical Engineering, John Carroll to a Professorship in Solid State Electronics, and Ken Bray to the Hopkinson and ICI Professorship of Applied Thermodynamics. In the same year, a Professorship of Management Studies was endowed by the accounting firm Peat, Marwick and Mitchell to which Stephen Watson was elected in 1985. The 'Shift to Science and Technology' scheme introduced by the government awarded three teaching and two assistant staff posts to the Department, and although three lecturers had retired or resigned, 11 new members were appointed, including Professors of Manufacturing Engineering, Aeronautical Engineering and Structural Mechanics.

Mike Ashby resigned his Chair in the Department upon his appointment as Royal Society Professor, creating a vacancy. Steve Williamson moved from Imperial College to fill the vacancy as Professor of Electrical Power. Alistair MacFarlane left to become Vice-Chancellor of Heriot-Watt University and Keith Glover succeeded him as Professor of Control Engineering. In 1989, a donation from Rolls-Royce established a Professorship of Aerothermal Technology, to which Nick Cumpsty was elected. It was an exceptionally busy

CONTINUING EDUCATION FOR INDUSTRY

Had he been alive, James Stuart, founder of the Engineering Department and powerful proponent of extramural studies in the University of Cambridge, would have been jumping with joy at the announcement by Heyman that 'following discussions with the Board of Extramural Studies, a small unit has been established in the Department to develop a programme of advanced short courses for professional people'.

The course was named the Cambridge Programme for Industry, and ran for the first time in 1989/90, with seven modules based on expertise within the Engineering Department and five based on expertise elsewhere in the University. External speakers from industry also participated in the courses, which lasted from two to six days. The following year, more than 20 new courses were added, with yet more in the planning stages. The key elements were to include industry specialists as teachers and to design courses for the open market as well as for a particular company or industry. It was asserted that continuing education is an important aspect of any company's investment in its employees, in new technology and in its management practices. In 1990/91, it was envisaged that courses would become increasingly popular with the academic staff despite their 'highly pressurised agenda of undergraduate and postgraduate teaching, and most importantly of research'.

Alec Broers, Heyman's successor, reported in the next year that the Cambridge Programme for Industry had been remarkably successful and needed to expand, and that it had therefore removed to its own premises close to the Engineering Department, No. 1 Trumpington Street. The programme, nurtured in the Engineering Department, was now an independent entity, but close collaboration continued. The programme evolved into the Cambridge Institute for Sustainability Leadership (CISL).

Stuart House, built in 1926, was for many years home to the Cambridge Board of Extramural Studies. It was named after James Stuart, the founder of the Engineering Department. After his death in 1913, his widow established an extramural lectureship in his memory and contributed to the cost of building Stuart House. Stuart House has been home to the Cambridge University Careers Service since 1976.

Below right: Heyman captured during a reflective moment.

period for a newly appointed Head of Department, but Heyman managed it with consummate skill, justifying all expectations.

Heyman spent the whole of his working life, 41 years, at the Engineering Department of the University of Cambridge as a University lecturer, reader, ad hominem professor in 1971, and finally as Head of Department for nine years before retiring in 1992. After retirement, Heyman continued as an engineering consultant and writer, and worked for a number of English cathedrals and churches. He also served as a member of the Cathedrals Fabric Commission for England and on the Architectural Advisory Panel for Westminster Abbey. In 1971, he was responsible for the restoration of Ely Cathedral's Great West Tower, and the immense value of his work was recognised by the Archbishop of Canterbury with the award of the Cross of St Augustine. He continued writing into his nineties and published more than a dozen books on his research and analysis of masonry structures, with particular reference to cathedrals and churches.

CHAPTER SEVEN

Into the Twenty-First Century

Alec Broers (b. 1938)

Professor of Electrical Engineering (1984–1996);

Head of Engineering (1992–1996);

Vice-Chancellor of the University of Cambridge (1996–2000)

There could be only one possible successor to Heyman – Alec Broers, waiting in the wings with outstanding academic and management credentials, was immediately elected 10th Head of the Engineering Department. Broers returned to Cambridge after two decades in the USA working for IBM, where he gained a wealth of management experience, rising to a high level of authority and eventually to the privileged status of IBM Fellow. He was appointed to the Chair of Electrical Engineering and the headship of the Electrical Division of the Engineering Department in 1984 and shortly afterwards his academic credentials were confirmed, being elected to Fellowships of the Royal Society and the Royal Academy of Engineering, and to the Mastership of Churchill College (1990–1996). The academic staff consulted by the Chairman of the Faculty Board could not conceive of a better qualified candidate for the position of Head of Department.

As a matter of record, Heyman had diffidently indicated his willingness to step aside in favour of Broers as Head of Department when Shercliff died prematurely in 1983, but Broers, still living in the USA, had declined, explaining that he would need to understand and settle into academic life, become accustomed to living on a fraction of his earnings at IBM, and learn to resist the temptation of returning to the USA!

Broers moved to Cambridge as Professor of Electrical Engineering in 1984. Immediately, with his characteristic energy, he raised funds for the clean room he needed for his research, and acquired semiconductor fabrication equipment from industry. The clean room was completed in 1987 and the centrepiece was a 400 kV scanning transmission electron microscope designed for the fabrication of nanostructures. With this and other machines, he and his team extended the technology of miniaturisation to the atomic scale, and the laboratory was named the Cambridge Nanofabrication Facility. In 1986, Broers delivered the

Alec Broers played a significant role in taking the Cambridge high-technology cluster to a new level. In 2006, he said: 'My model for partnership between industry and academia is one where collaborative groups are formed that work jointly … universities have to work with industry if they want to do effective research.'

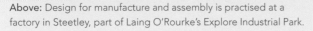

Above: Design for manufacture and assembly is practised at a factory in Steetley, part of Laing O'Rourke's Explore Industrial Park.

Below: The Millennium Bridge in London. The bridge opened in 2000, but was closed after three days as it began swaying due to a subtle interplay of the footfall of hundreds of people walking together and the dynamics of the structure. David Newland, Head of Department from 1992 to 2002, advised the London Millennium Bridge Trust on the remedial work. Modifications involved the installation of 91 dampers, similar to car shock absorbers, designed to stop the oscillations.

Above: VAAC Harrier on which control laws developed by Keith Glover with Rick Hyde and George Papageorgiou were flown with success.

Below: Examination of samples using a scanning electron microscope.

Top: The Broers Building, part of the Hauser Forum on the West Cambridge site, opened in 2010. It provides modern office space and resources to small- to medium-sized enterprises, which benefit from working in partnership with academic teams and other companies in the Cambridge high-technology cluster.

PUTTING ENGINEERING AT THE FOREFRONT: THE SCHOOL OF TECHNOLOGY

While attending meetings of the Council of the School of Physical Sciences, the University body charged with advising the General Board on the distribution of resources in a fair and equitable manner, Alec Broers and Jacques Heyman came to the conclusion that the needs of technological subjects demanded more time and attention within the Council's deliberations and decisions.

Broers argued for an independent School of Technology separated from the somewhat cumbersome School of Physical Sciences, which had 16 members representing a very wide constituency of faculty interests. The University accepted his proposal for a new school that would emphasise the importance of engineering and cognate technological subjects in the University.

The School of Technology, founded in 1993, was composed of four members: Chemical Engineering, Computing, Engineering and the Judge Institute of Management. The independent School was thereby qualified to have its own representative on the General Board of the University and thus a stronger voice on matters affecting technology and University policy issues; the creation of a School of Technology was a particularly notable achievement by Broers in his tenure as Head of the Department. (By 2016, the Institutions within the School were: the Department of Chemical Engineering and Biotechnology, the Computer Laboratory, the Department of Engineering, the Judge Business School and the Cambridge Institute for Sustainability Leadership.)

Below: Broers worked for IBM at the Thomas J Watson Research Center in the United States for 19 years before joining the Department of Engineering in 1984.

Clifford Patterson Lecture 'Limits of Thin-Film Microfabrication', which was published in the *Proceedings of the Royal Society*.

Under Broers' guidance, the Department, still dogged by lack of space for its ambitions, decided to carry out a long-term review of its needs. It came to the inevitable conclusion that the Scroope House Estate had been exploited to its maximum capacity, and that it was now totally inadequate for the Department's anticipated future. After all, Baker had written in the 1950s that he had built enough research space for half a century but no more! And long before Baker, Inglis had noted that at some time in the future the Department would have to acquire Scroope Terrace to gain more space and to build a magnificent colonnaded entrance facing onto Trumpington Street.

The review noted that the Department was now growing faster than the University as a whole, and that at this rate of growth its total numbers would reach 4,000 by the year 2020. The planned four-year course would alone

expand the Department by 20 per cent, which could not be accommodated with any conceivable development of the Scroope House Estate. The analysis of the space needed for research also confirmed that development of the present site was not a viable option, and it was concluded that the Department should move to a greenfield site in West Cambridge by 2010 at the latest – wishful thinking, as the Department is still centred where it was in 1993, although again planning with determination to move to West Cambridge.

DEVELOPMENTS IN RESEARCH AND ACADEMICS

In his first year as Head of Department, Broers reported that the Department's research was stronger than ever before, with a 5A rating, the highest awarded, in the Higher Education Funding Council's Research Assessment Exercise (RAE), and he also noted that the Department was the largest 'general engineering' department in the country, with 135 academic staff members who were, without exception, sufficiently productive in research to be included for consideration in the RAE. Research continued to grow rapidly during Broers' four-year tenure, with 2,000 research papers published and almost 300 PhD degrees awarded.

The four-year Tripos, developed under Heyman, was also at last ready to be introduced, after many years of planning and much hard work by overstretched academic staff. The aims of the new courses were set out by Broers as follows:

LANGUAGE SKILLS FOR THE INTERNATIONAL ENGINEER

The globalisation of the world economy, increasing international competition as well as cooperation across national boundaries and, particularly for the UK, pan-European competition and collaboration had created a need for engineers with a global outlook. With a remarkable leap of faith it was concluded in the Engineering Department that, as Broers wrote: 'Learning foreign languages [must] therefore become an integral part of the new four-year engineering course.'

Sarah Springman and Ann Dowling led the effort to raise funds for teaching languages and a Language Laboratory was established in the Department by adding a mezzanine floor to the Centre Wing of the Baker Building. The project was supported financially by a number of national and European bodies, and lectors in French and German were soon appointed. The Laboratory was formally opened by the Chancellor, His Royal Highness The Duke of Edinburgh, in 1994.

Optional courses for first- and second-year students were well attended, and third-year undergraduates were encouraged to present their work, if they wished, in either French or German! The initiative won the Languages for Export Award from the Department of Trade and Industry.

Above left: The Language Laboratory in use today.

Microsoft President, Bill Gates, centre, on a visit to Cambridge in 1997. Microsoft chose Cambridge as the location for its first research lab outside the United States and appointed Roger Needham (on the left), the distinguished engineer and computer scientist, as its head. On the front right is Alec Broers, who was then Vice-Chancellor of the University.

to maintain a broad scientific base on which to construct the principles of engineering; to accommodate students less well prepared by schools in mathematics and physics; to prepare students for future participation in European industry; to place more emphasis on communication skills and on design projects; to aim to relieve some of the pressure on students; and finally to promote European exchanges. It was also decided that some expertise in a foreign language would be an integral part of the four-year engineering course.

The Department also confirmed that engineering design would be emphasised in the new Tripos, and a good example had been set by designing full-custom CMOS (complementary metal-oxide semiconductor) integrated circuits in the Department before sending them off for fabrication in a silicon foundry; the chips were returned and tested in the Department to confirm the success of the designs. It was planned to allocate more resources to information engineering and to install a modern computer teaching facility with 100 workstations operating simultaneously.

The first two years of the four-year Tripos went smoothly, and opinion surveys showed that most students had reacted positively towards the courses. In the academic year 1994/95, the innovative third year of the course was instituted, with traditional laboratory experiments reinforced by a 'Professional Group Activity', which gave students the opportunity to experience the type of challenge associated with their chosen area of specialisation. To accommodate the needs of the new Tripos, examinations were held at the beginning of the Easter term rather than at the end, followed by two four-week projects. Undergraduates would then spend half of their fourth year on a major project that would, wherever possible, be linked to an industrial partner. It was claimed that benefits would arise through this linkage for the student, the industrial partner and the academic staff.

Cambridge Foundation Teaching Prizes (later known as Pilkington Prizes), founded to recognise 'Excellence in Teaching', were awarded to Ken Wallace, Head of the Engineering Design Centre, in 1994, and to Robin Porter Goff, lecturer in design, in 1995. These prizes are highly valued throughout the University.

Mike Gregory retired as Head of the Manufacturing and Management Division and of the Institute for Manufacturing (IfM) in 2015. He was the founder member of the team which established the IfM in 1998, linking science, engineering, management and policy and integrating education, research and practice.

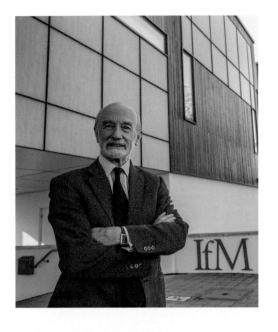

CHANGES IN FACULTY

Bill Crossland was appointed the Northern Telecom Research Professor of Photonics in 1992, and in the next year Ann Dowling became the first female professor in the Department. In 1993, the tragic and sudden death of Frank Fallside, Professor of Information Engineering, led to the naming of a new laboratory, on the fourth floor of the Baker Building North Wing, in his honour. The Fallside Laboratory provided a new home for speech and vision research and control engineering. Fallside was a pioneering contributor to the field and his work established the subjects in the mainstream of engineering teaching and research. His legacy continues to thrive in the Department through his laboratory, which is the home of machine intelligence research.

Mike Gregory was elected to the Chair of Manufacturing Engineering and appointed Head of the Manufacturing Engineering Division in 1995. A year earlier, Steve Young was appointed Professor in the Information Engineering Division headed by Professor Keith Glover, and Nick Cumpsty and Steve Williamson were elected to Fellowships of the Royal Academy of Engineering in 1995. Broers announced a donation by Hamid Jafar for a Chair in Petroleum Engineering and Andrew Palmer was appointed in 1996.

As for Broers, his term as Head of Department was to be brief, since he was elevated to the position of Vice-Chancellor of the University of Cambridge in 1996. He was the first engineer to be appointed to this position – a singular honour for him and for the Engineering Department.

FRANK FALLSIDE

LECTURER, READER AND PROFESSOR

OF

INFORMATION ENGINEERING

1961-1993

Left: Andrew Palmer was appointed to the Jafar Research Professorship of Petroleum Engineering in 1996. He retired from Cambridge in 2005, but continued at the forefront of his field. Here he is shown in 2007, when Keppel Offshore & Marine Limited and the National University of Singapore (NUS) appointed him to the Keppel Chair at NUS, where he had been a visiting professor.

Above: Plaque honouring Frank Fallside at the Machine Intelligence Laboratory.

David Newland (b. 1936)

1875 Professor of Engineering (1976–2003)

Head of Engineering (1996–2002)

The unprecedented appointment of the Head of the Engineering Department, Alec Broers, to the Vice-Chancellorship of the University of Cambridge, and his declared intention of leaving the Department on 31st March 1996 to prepare himself for his new position, created an immediate vacancy for the headship of the Department, and a Search Committee under the Chairman of the Faculty Board, David Harrison, was given the remit of identifying a suitable successor. Soundings taken among the academic staff and responses identified strong support for David Newland, who was duly appointed Head of Department for a period of six years from 1st October 1996. He also served as acting head from 1st April 1996, while Broers was on leave.

Newland stood out among his peers for his significant contributions to the Department in setting up the Engineering Design Centre as well as the Manufacturing Engineering Tripos, and for his management of the Department's Division of Mechanics, Materials and Design. His wide experience as a consultant to industry in the UK and the USA was highly regarded, as were his authoritative contributions at the Flixborough chemical plant disaster inquiry in 1974. His first task was to send the Department's warmest congratulations and good wishes for his future role to the Vice-Chancellor elect. He also acted as one of the Deputy Vice-Chancellors of the University of Cambridge (1999–2003) while serving as Head of Department.

The first year of Newland's headship of the Department was marked by two events: the completion by the first cohort of students of the fourth year of the Engineering Tripos and the Electrical and Information Sciences Tripos and therefore the very first awards of MEng degrees in the University of Cambridge. By all accounts, the new Tripos had been a great success with the undergraduates, and was commended by external observers. The academic staff members who had dedicated so much time and energy to its preparation and delivery were extremely satisfied by this outcome. The range of tasks undertaken by the students, the major project in the fourth year, flexibility in the taught modules and the well-designed laboratory work were all well received and much appreciated. The design of the four-year Tripos allowed for new styles of teaching and assessment, offered a range of choices to undergraduates, and created many avenues for specialisation. The four-year Engineering Tripos had changed the three-year

David Newland on the south side of the Millennium Bridge. His research has focused on engineering design and, in particular, on vibration analysis and control in engineering design.

The Flixborough disaster involved an explosion at a chemical plant close to the village of Flixborough in North Lincolnshire in 1974. Twenty-eight people died and 36 were seriously injured. Newland was an expert witness at the inquiry, which found numerous failings in the running of the plant and concluded that the 'disaster was caused by the introduction, into a well-designed and constructed plant, of a modification which destroyed its integrity'.

Tripos out of all recognition and given a decidedly modern, twenty-first-century look to Cambridge engineering in anticipation of the new millennium. This also brought the Cambridge Engineering Tripos more in line with undergraduate courses within Europe.

The Department's proposal for a reorganisation of its management structure was approved by the University and three Deputy Heads were appointed: Martin Cowley, Deputy Head for Teaching; Keith Glover, Deputy Head for Research; and Ann Dowling, Deputy Head of Graduate Studies. Glover was also appointed chair of a working party 'Exploring Research Strategy'. Steve Williamson took over as Chairman of the Space Committee, which progressively refurbished and modernised accommodation on the Scroope House Estate.

The idea of moving research, but not initially teaching, to West Cambridge gained strength and plans were laid. Not all members of the Department approved of these plans, but Newland's view was that they should be encouraged, with the first phase to be the removal of the Institute for Manufacturing, followed by Electrical Engineering, and that two-site working should be tolerated for as long as it took to transfer the whole department to West Cambridge – this became the Department's aim. During this time, the Whittle Laboratory at West Cambridge was extended, with the extension being opened by John Cadogan, Director-General of the Research Councils.

Another innovation in this period was the biannual publication of the joint newsletter of the Cambridge University Engineering Department and the Engineers' Association, with the first issue appearing in the summer of 1993. The purpose of the newsletter, called *Enginuity*, was to keep all members of the Department, past and present, informed of the research in the Department as well as the affairs of the Association, replacing the annual report sent to members of the Association in the past. A very wide range of research was represented in 10 years of somewhat irregular publication before *Enginuity* was replaced by the biannual *Newsletter*, first published in 2005. *Enginuity*'s circulation was measured in hundreds; the current circulation of the Department of Engineering *Newsletter* exceeds 20,000.

Enginuity (left) was first published in 1993. It was replaced by the Department's biannual *Department of Engineering News* (right).

Pages from *Enginuity* 1, 1993 (left) and *Enginuity* 6, 1997 (right).

In 1995, five ROPA (Realising Our Potential Awards) were gained, rewarding research workers who were already receiving significant research funding from industry and enabling them to make more proposals for blue sky research projects. In January 1998, a Teaching Quality Assessment of the Department was carried out by a team of 15 assessors appointed by the government. Extensive preparations for this week-long visit were necessary, and staff efforts, led by Martin Cowley and Ken Wallace, were amply rewarded when it was announced that the Department had received the outstanding result of 23 points out of a maximum of 24.

A major refurbishment and expansion of the Department Library was completed in 1998, providing much-needed additional seating accommodation for students as well as more bookshelf space. A Design Centre was created in the space above Lecture Room 3 and substantial space for research became available by reconstructing the roof and creating a loft conversion. The new and much improved frontage is best viewed from Coe Fen.

The Royal Academy of Engineering held its summer soirée in the Department on 13th July 2000. Next day there was a Reunion and Conversazione of the Cambridge University Engineers' Association (CUEA) to mark the 125th Anniversary of the Department. To support these events, the Department arranged an exhibition of its research and industrial collaboration, with exhibits spreading into the Scholars' Garden of Peterhouse, so extensive were the displays. Distinguished guests included HRH The Duke of Edinburgh (who had opened the Baker Building in 1952) and the CUEA President, Baroness Platt of Writtle.

RESEARCH UNDER NEWLAND

If Baker had been alive at the turn of the millennium, more than half a century after his appointment as Head of Department, he might have thought he was witnessing the final realisation of his dream of transforming the Engineering Department into a centre of research of international standing, but he might well have been taken aback to discover that Newland's Department was breaking traditional boundaries between disciplines and directing itself towards addressing the world's most pressing problems with engineering solutions. These ambitions were apparent in Newland's detailed account of Department-wide research in the submission he prepared, with the help of his senior colleagues, for the adjudicators of the Research Assessment Exercise (RAE) in 2000/01.

In his submission, Newland outlined the Department's research plans for the future in 15 subject areas, pressing the point that 'The advantages of being an integrated engineering department include encouraging *cross-disciplinary collaboration* and we envisage continued and growing overlap between our research groups.' He stressed that collaboration between different groups is found across the whole range of engineering disciplines that are

Front cover of the publication produced to commemorate 125 years of the Engineering Department.

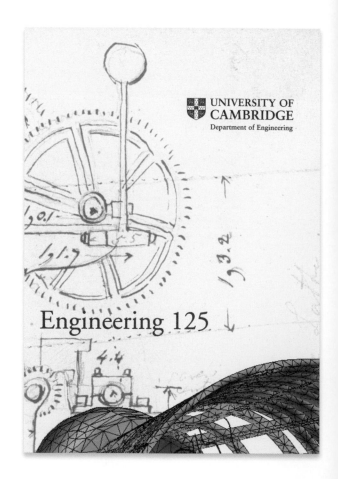

Peter Guthrie became the first Professor in Engineering for Sustainable Development in the UK in 2000.

researched in the Department, and gave examples of collaborative projects. Funding from both industry and the traditional sources of government Research Councils was cited and the appointments of Professors Hopper, Mair and Guthrie described as catalysts for initiatives and expanded experimental facilities.

Newland also pointed out some special events in this period, such as the valuable support of Rolls-Royce, leading to the establishment of a University-wide Gas Turbine Partnership with Ann Dowling as its first Director. Mark Welland was appointed Director of the Interdisciplinary Research Centre in Nanotechnology, funded by an exceptionally large grant from the Research Councils which included a purpose-built laboratory in West Cambridge. Newland added that the Strategic Research Infrastructure Fund had made it possible to extend the Schofield Centrifuge Centre and to improve its experimental facilities.

After more than 20 years at Cambridge University Computer Laboratory, Andy Hopper was elected Chair of Communications Engineering in the Engineering Department in 1997. He returned to the Computer Laboratory as Head of Department in 2004 and the two departments continue to work collaboratively.

THE COMPETITION FOR PRESTIGE AND FUNDING IN THE RESEARCH ASSESSMENT EXERCISE

The government, or more specifically the Higher Education Funding Council for England (HEFCE), decided in 1992 to evaluate the quality of research in universities in the UK by means of a rigorous process, the Research Assessment Exercise (RAE). It was made clear to all universities that a poor performance in the RAE would incur financial penalties, and that bad publicity could follow for an underperforming university. In 2001/02, the Engineering Department was required to assemble detailed information about its operations and to provide statistical data on research over the previous four years. A statement accompanied by a list of their four best publications was required from each member of the Department; a demanding amount of work was placed on the shoulders of Divisional Heads and the Director of Research, Malcolm Macleod.

Newland wrote that 'The Department presented itself in 15 research sub-areas each headed by a professor', and every effort was made to demonstrate that the work of the Department not only met the highest of national standards, but that most areas were of international eminence. To the immense gratification of the Head of Department and all his colleagues, the Department was rated at the very top grade, 5*A. The five stars signified the highest research rating and the letter A denoted that more than 95% of eligible staff had been included in the exercise. In other words, the Department was not only reaching the highest quality in research, but also almost all of the academic staff members were active research workers. This outcome was a major landmark in Newland's successful leadership of the Department over the millennium, 1996–2002.

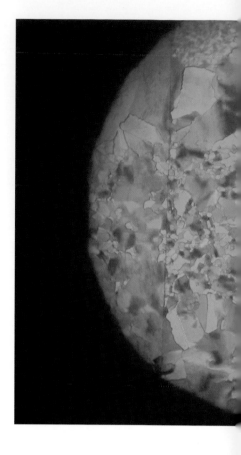

CHANGES IN FACULTY

The Department's academic strength in this period was demonstrated by spectacular growth in the professoriate. Robert Mair, John Young and Gehan Amaratunga were elected to established professorships, while Andy Hopper and Robin Langley were elected to newly-created ones. Ian White was elected to the endowed van Eck Professorship and Michael Kelly to the Prince Philip Professorship. Andrew Palmer, Bill Crossland, Simone Hochgreb, Harry Coles, Howard Hodson and Ian Hutchings were all elected to industrially-funded professorships. Peter Guthrie was appointed to the Royal Academy of Engineering Research Professorship for Sustainable Development. And there were 14 personal promotions to professorships.

In retirement, Newland has had more time for his hobby of visiting places where he can find butterflies flying wild in their natural habitats, and has published three books on the ecology of butterflies and moths. These include *Discover Butterflies in Britain,* written in 2006, with delightful illustrations and short poems by his wife adding to the pleasure of reading the book.

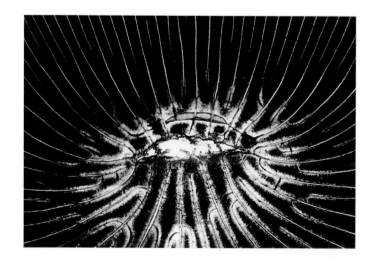

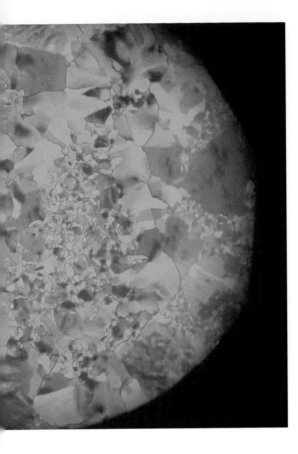

A PAIR OF ENNOBLED PRESIDENTIAL CASTAWAYS

Among the many academics from the Department feted with public honours and prestigious awards, Alec Broers and Ann Dowling have been prominent. They are unique in the Department's history having hit all of the following high notes.

- Broers and Dowling have both been Head of the Department of Engineering (1992–1996 and 2009–2014 respectively).
- They both served as the President of the Royal Academy of Engineering: Broers from 2001 to 2006 and Dowling from 2014, when she became the first woman to hold the presidency.
- In 2004, Broers was made a life peer for his contribution to engineering and higher education and in 2007 Dowling was recognised as a Dame in the New Year's Honours List for her services to science. She was appointed to the Order of Merit in 2016.
- On a lighter note, both Broers and Dowling have been 'castaways' on Radio 4's *Desert Island Discs*. Broers in 2001 recalled his arrival in Cambridge as a Choral Scholar at Caius College and said of his work: 'If cars had made the same progress as electronics in the past decade, then you would be able to drive from Cambridge to London in half a second.' He chose 'lots and lots of chocolate' as his luxury item. Dowling in 2016 spoke to Kirsty Young, the host, about her fascination from a young age of working out how things worked – taking her bike apart and the lights of her dolls' house aged six. She chose the *Collected Papers of Sir James Lighthill*, the applied mathematician, polymath and pioneer in the field of aeroacoustics, as her extra reading material and an artist's studio with paper, pencils and watercolours as her luxury.

Above: Blue phase liquid crystals self-assemble into building blocks that form three-dimensional structures. The size of the blocks can be tuned with chemistry and applied electric fields to match wavelengths of visible light. The material has the potential for use in display and communication technology. The colourful image was taken using a polarising light microscope.

Opposite: Optical microscopy image showing the drying and cracking of a film formed when an alkyl ketene dimer dispersion is deposited and dried on a glass microscope slide. The experiment was conducted in the Inkjet Research Centre, which works closely with a consortium of companies to explore the fundamental science and engineering of inkjet technology.

Right: Coal power plants are the major sources of CO_2 emissions. Simone Hochgreb and her team research ways of reducing the pollutant emissions by burning the coal particles in an oxyfuel environment rather than normal air.

Keith Glover (b. 1946)

1974 Professor of Engineering (1989–2013);
Head of Engineering (2002–2009)

David Harrison, Chairman of the Faculty Board, was appointed in 2001 to chair a sub-committee of the Board, with the remit of selecting a new Head of Department from 24 of the professors in the Department. On the basis of both written statements and soundings taken among the academic staff, the preferred candidate was Professor Keith Glover.

Glover had joined the Department as a University lecturer in 1976 after seven years in the USA, obtaining his PhD from MIT and being an Assistant Professor at the University of Southern California. His academic credentials in control engineering were self-evident, given that he was a Fellow of both the Royal Society (1993) and the Royal Academy of Engineering (2000). In terms of administrative experience and management responsibility, he was the Head of the Information Engineering Division, Deputy Head of the Department for Research, Chairman of the Council of the School of Technology and Member of the General Board. What more could one ask? The only question was his willingness to take on the task, considering his commitment to research and the reluctance he had expressed in his written submission to the sub-committee.

On receiving an invitation from Harrison asking him to take on the headship of the Department, Glover responded by writing: 'I'm prepared to be nominated but do so with substantial reservations because I believe that the expectations from the Head of Department, in this next critical phase of the Department's development, need to be realistic.' The demands on the Head of Department had indeed grown with each successive incumbent. Research activity was considerably more extensive and the task of dividing the limited resources, particularly space for research projects, in a fair and equitable manner was not easy, and was the cause of frequent acrimony in the Department. Not everybody knew that Glover's cautious and unassuming manner hid a steely determination and great strength of purpose – after all he had been a 1st Dan black belt in judo and had captained the University of London's judo team as an undergraduate! In the event, he accepted the position, becoming the first Head of Department known to have been educated in a state school and the first since Ewing not to have been an undergraduate at the University of Cambridge.

Perhaps the most important action by Glover as Head was the first comprehensive strategic review of the Department's aims and objectives and the preparation of a development plan. Glover based the review on in-depth consultations with academic staff and external stakeholders in the Department,

Keith Glover's research interests include feedback systems, robust control and model approximation. His theoretical work has been tested and developed in applications in aerospace and automotive engine management.

Glover was a judo enthusiast as a student at Imperial College and captained the University of London judo team. This image from the magazine *Judo* shows him scoring in national under-21 trials.

with the aim of developing a rigorous and forward-looking strategy for research, infrastructure, teaching and administration. The preamble to the review, coordinated by Philip Guildford, newly appointed Director of Research, stated that 'the aim of the Department is to address the world's most pressing challenges with science and technology, working in collaboration with other disciplines, other institutions, commercial companies and the entrepreneurial community'.

Although the Engineering Department was founded to teach engineering for the BA degree as an integrated discipline and maintained this philosophy throughout its history, this amalgamation of conventional engineering sub-divisions did not apply to research. Following the strategy review, the level of assimilation of separate research topics was significantly enhanced; cross-linking themes were developed within Departmental research and connections were established with subjects ordinarily unrelated to engineering. The strategy review observed that the Department's buildings had been steadily dispersed across Cambridge to meet immediate needs but despite this diversification many research areas remained unfit for purpose. It was decided to reinforce the policy of concentrating developments on the two 'sister sites' of the Scroope House Estate and University-owned land in West Cambridge.

Glover also took the innovative step of establishing an 'International Visiting Committee' of leading engineers from universities and industry, at home and abroad, to meet biennially to appraise the Department's performance and strategy and to offer guidance to the Department concerning its operational procedures as well as the plans for its future. The Founding Committee, chaired by Alec Broers, concluded in its report that 'the Department's strategy was well considered and expressed' and found the research to be of a high standard and relevant to industry's needs. It advised the Department to include more details of implementation and to track the progress with numerical data for the next strategic review.

On 23rd June 2003, the Cambridge University Engineers' Association celebrated the 50th anniversary of HRH The Duke of Edinburgh as its Patron. A sundial representing the Association's logo, designed by a member of the Engineering Department, Peter Long, and manufactured in the Department's workshop, was presented to the Chancellor at a reception in the gardens of the Vice-Chancellor's Lodge in Latham Road.

Above: The Electrical Engineering Division moved into spacious new premises in West Cambridge in 2006.

Right: Sundial presented to HRH The Duke of Edinburgh in 2003.

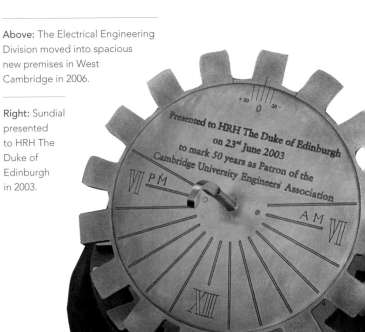

Below: The Schofield Centre for Geotechnical Process and Construction Modelling focuses on the key questions of construction and environmental technology, where field, computational and physical modelling studies can be integrated in collaboration with industry.

Right: An engine test cell viewed from the control room. From left to right: Glover (previous Head of Department), Paul Dickinson (postdoctoral researcher) and Nick Collings (previous Head of the Energy, Fluid Mechanics and Turbomachinery Division). Their work is a remarkable long-standing collaboration between control and combustion engineering.

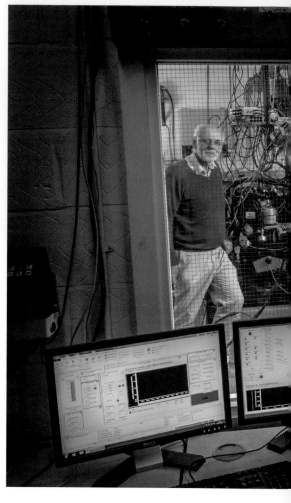

Sir Frank Whittle
OM, KBE, CB, FRS
1907 – 1996
Jet Propulsion Pioneer.

Pursued the development of his jet engine at the Cambridge University Department of Engineering.

The Whittle blue plaque at the entrance to the Department.

During Glover's tenure, there was an immense amount of building activity in West Cambridge on behalf of the Engineering Department, unparalleled since the days of Baker's development of the Scroope House Estate. The Schofield Centre for Geotechnical Process and Construction Modelling had its operational opening in 2002 and the Nanoscience Centre was opened by the Chancellor in 2003. The Electrical Engineering Division moved into a striking and spacious new building in 2006, which provided the division not only with much-needed space but also added much improved experimental facilities, including state-of-the-art clean rooms. The Department's Institute for Manufacturing moved from its cramped basement space in the old University Press site to the splendid Alan Reece Building in 2009. The centre of gravity of the Engineering Department was shifting decidedly westwards during Glover's tenure, with almost 50 per cent of the research activity based there. The removal of Electrical Engineering to its new home at West Cambridge vacated a large amount of space on the Scroope House Estate, which was redeveloped to house the new areas of Engineering Design, Bioengineering, Computational and Biological Learning and Sustainability.

A blue plaque in honour of Sir Frank Whittle was unveiled at the main entrance to the Department on 13th January 2004 at an event organised by

A DYNAMIC ENGINEERING DEPARTMENT (2002–09)

The vigorous growth of the Department is best illustrated by noting that, during a six-year period in Glover's seven years of headship, roughly 5,000 conference and journal papers were published, 82 books were published and more than 100 patents were taken out by members of the Department. Approximately 500 PhD degrees were awarded. Six professors and 33 University lecturers were appointed, and there were 15 ad hominem promotions to chairs and 21 to readerships – an unprecedented rate of appointments and promotions.

Ten members of the academic staff were elected to Fellowships of the Royal Academy of Engineering, making 16 in all, and four were elected to Fellowships of the Royal Society, making six in all. A DBE for services to science was announced in the New Year's Honours list for Dowling. There were other awards to academic staff: Ffowcs Williams received the Sir Frank Whittle Medal, John Denton the James Clayton Prize, Ken Johnson received the Royal Medal of the Royal Society and Ken Bray was awarded the Sugden Prize.

Tom Smith training Indian partners' staff in the commissioning and operation of a solar-powered NIFTE pump during a Thermofluidics training visit to Pune, India, in May 2016. Thermofluidics is conducting field trials in India, Bangladesh and Kenya.

the Sir Arthur Marshall Institute for Aeronautics, and lectures on 'Future Prospects for Aeroengines' and 'Green Aviation' were held in the Upper Hall of Jesus College.

The Times organised a 'One Minute Pitch' competition in 2005 with a prize of £100,000 and the Department was delighted when Tom Smith, a PhD student, won the prize with the description of his development of a cheap and efficient pump with no moving parts, designed to help farmers in the developing world irrigate their crops. The judges were unanimous in choosing his entry from the many thousands submitted in the competition.

History was made when an alumnus of the Engineering Department, Mohan Munasinghe, was awarded the Nobel Prize for Peace in 2007. As Vice-Chair of the United Nations Intergovernmental Panel on Climate Change (IPCC), he shared the Prize jointly with Al Gore and IPCC colleagues. Munasinghe read for the Mechanical Sciences Tripos, graduating with BA (Hons) in 1967. His work has focused on energy, water resources, sustainable development and climate change for the last 50 years.

In October 2007, 'Engineering for the Life Sciences' became a full 'engineering area', equal in status to Mechanical or Electrical and Information Engineering for Part II of the Tripos. The development followed the appointment of Daniel Wolpert in 2005 to the Foundation Professorship of Engineering (1875), with the specific remit of developing Engineering for the Life Sciences, and subsequently five lecturers with interests in this area were appointed.

The 10-metre beam centrifuge at the Schofield Centre is used to study earthquakes, foundations, piling, tunnelling and retaining walls. It can take one-tonne experiments to 100 times the acceleration of gravity. The team constructs models of civil engineering projects and spins them in the centrifuge so that the forces acting on these models scale correctly.

A new type of 'integrated coursework' for second-year students was also developed, in which a single physical system was studied within a number of differing academic disciplines. As an example of integrated coursework, a study was centred on a model of an earthquake-resistant building. Undergraduates investigated the structure of the building, its dynamics and soil mechanics relating to its foundations, and developed techniques for the measurement and analysis of data from signals transmitted by an instrumented model. They were also given the opportunity to plan and perform their own experimental work with the apparatus. The pedagogical development of this coursework was part of a wider educational collaboration within the University, the Teaching for Learning Network, coordinated jointly by the Department of Plant Sciences and the Centre for Applied Research in Educational Technologies.

RESEARCH UNDER GLOVER

The Department continued to grow rapidly, and by 2005 the number of academic staff had risen to 132, contract research staff to almost 200 and research student numbers were approaching 600, with the annual budget running to several tens of millions of pounds. Research activity grew in line with these increasing numbers. The high international status of the Department was accepted throughout the University as well as nationally, and this standing was amply confirmed by high positions gained by the Department in the University ranking systems initiated by Shanghai Jiao Tong University and the *Times Higher Education Supplement*.

The best source for detailed information on the wide range of research and the many exciting outcomes from the research projects as well as events in the Department, ranging from new grants and donations to the appointment of new teaching staff, is the Department's *Newsletter*, which was published biannually during Glover's tenure.

The research aims of the six academic divisions are summarised in the box opposite, extracted from the Department's submission to the 2008 Research Assessment Exercise, under which it was ranked at the highest level and dramatically ahead of any other department in its area. The cross-linking themes showcase new developments within the Engineering Department.

Above: Mohan Munasinghe, winner of the 2007 Nobel Peace Prize as part of the Intergovernmental Panel on Climate Change, studied engineering at Cambridge and graduated in 1967.

Opposite: Norman Fleck, noted for his work on micromechanics, the properties of materials, and novel materials such as metal foams.

RESEARCH AMBITIONS OF THE SIX ACADEMIC DIVISIONS IN 2008

Ann Dowling, Energy, Fluid Dynamics and Turbomachinery: research into fluid mechanics and thermodynamics, aimed at mitigating the environmental impact of ground and air transport.

Bill Milne, Electrical Engineering: research into advanced materials, nanotechnology, sensors, energy generation and conversion, display and communication technologies.

Norman Fleck, Mechanical Engineering, Materials and Design: research into micromechanics and materials, design and dynamics and vibrations and using cross-disciplinary partnerships working on bioengineering and healthcare systems.

Robert Mair, Civil, Structural and Environmental Engineering: research into design, construction and operation of sustainable infrastructure designed to husband the earth's resources and to protect the environment.

Mike Gregory, Manufacturing Engineering: research into manufacturing technology, operations, strategy and policy, in close partnership with industry, in order to increase industrial competitiveness.

Steve Young, Information Engineering: research into fundamental theory and applications relating to the generation, distribution, analysis and use of information in engineering and biological systems.

CROSS-LINKING MULTIDISCIPLINARY THEMES, 2008

Engineering for Life Sciences (Leader: Daniel Wolpert): builds a quantitative mechanistic understanding of biological systems and hence seeks applications in biological and medical areas, and biologically inspired solutions in engineering.

Cognitive Systems Engineering (Leader: Steve Young): constructs integrated computer-based systems with cognitive qualities such as perception, learning, reasoning, decision-making, communication and action.

Sustainable Development (Leader: Peter Guthrie): integrates economic, social and environmental factors into sustainability and builds these processes into engineering design.

Ann Dowling (b. 1952)

Professor of Mechanical Engineering (1998–)

Head of Engineering (2009–2014)

After serving for seven years as Head of Department, Glover let it be known that he wished to stand down, and a sub-committee nominated by the Faculty Board was appointed to seek a successor through a transparent process. Its first task was to identify professors willing to be considered for the headship; just three were prepared to be nominated. These candidates were asked for position papers outlining their understanding of and vision for the Department, and to make presentations at open meetings. The outcome of this process was the unequivocal selection of Ann Dowling in the most open and public appointment to the headship since the Department was founded in 1875.

Dowling is a distinguished scholar whose outstanding work on unstable combustion in aeronautics led to a fundamental understanding of the mechanisms that cause jet engine instability. In recognition of this work, she was elected a Fellow of the Royal Academy of Engineering in 1996 and a Fellow of the Royal Society in 2003. During her 32 years in the Department, she served on a large number of academic committees, held the responsible positions of Deputy Head for both Teaching and Graduate Studies, and led one of the Department's six academic divisions; she was certainly well qualified to lead the Department. The appointment was widely reported, as she was the first woman to become Head of the Engineering Department – which had come a long way from the days when women were not admitted to the staff common room!

More than any other Head of the Engineering Department, Dowling was able to bring engineering into public prominence and to the notice of all sectors of society – including many where engineering was a mystery. In 2011, she received the Inspiration and Leadership in Academia and Research Award, and again her achievements were widely reported. Her contributions where engineering was creating a better environment for society, such as the silent aircraft initiative, were particularly commended. Dowling's achievements as an engineer came to the attention of a much wider audience when she was interviewed by BBC Radio 4 for its programme *The Life Scientific*, and even more so when she was nominated for the BBC *Woman's Hour* Power List 2013.

Ann Dowling's research is primarily in the fields of combustion, acoustics and vibration and is aimed at low-emission combustion and quiet vehicles.

Women in Engineering is one of the Department's most important initiatives. Claire Barlow (right), Deputy Head of Department (Teaching) and University Senior Lecturer, is a prominent figure in this campaign and is shown in dialogue with students.

Concept illustration of a silent aircraft developed by Cambridge researchers under Dowling along with the Massachusetts Institute of Technology. Adaptations for quieter operation include a single flying wing, top-mounted engines and the elimination of flaps and slats.

Dowling's pre-eminence within the national engineering profession was confirmed in 2014 when she was elected President of the Royal Academy of Engineering, in which capacity she was able to make the point that 'there is a growing recognition of the vital importance of engineering in addressing the many challenges that face society'. In this very significant new role she is in a position to influence government policy on engineering at the highest level and 'to bring engineering to the heart of society' – the motto adopted by the Royal Academy of Engineering.

In 2004, she chaired the report *Nanoscience and Nanotechnologies: Opportunities and Uncertainties*, which was produced jointly by the Royal Academy of Engineering and the Royal Society. It was a perfect example of how such a committee should act to deliver the best possible outcome. Since the beginning of her term as President of the Royal Academy of Engineering, the quality of reportage has improved significantly. Throughout her career, Dowling has shown a remarkable ability to serve the cause of engineering on Departmental and national committees. She was a Council Member of the Engineering and Physical Sciences Research Council (EPSRC) and outside the academic world she sits on the Board of BP as a non-executive director, a position of responsibility in the affairs of the sixth-largest oil and gas multinational in the world.

Right: The cover of MacKay's critically acclaimed book. 'I am not pro-wind or pro-nuclear: I am just pro-arithmetic. We need an energy plan that adds up. It is not going to be easy, but it is possible.'

Below: David MacKay (1967–2016), appointed the first Regius Professor of Engineering in 2013 and a Knight Bachelor for services to scientific advice in government and science outreach in 2016.

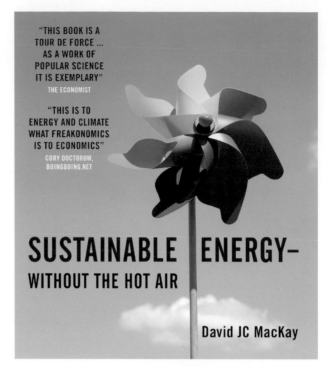

One of the landmarks of this era was the creation of the Regius Professorship of Engineering at Cambridge. Regius Professorships are Royal academic titles, created by the monarch. The Engineering role was announced in 2011 to celebrate The Duke of Edinburgh's 34 years as Chancellor of the University. The new post is designed to give an outstanding academic the opportunity to build on the Department of Engineering's world-leading research in fields that address major global challenges. These include: creating lasting energy solutions, building cities in the future, managing risks and driving innovation.

David MacKay was appointed as the first Regius Professor of Engineering in 2013. An eminent researcher in machine learning and information theory, and a Fellow of the Royal Society, MacKay is perhaps better known to the public for his groundbreaking work on sustainable energy and, in particular, as the author of the critically acclaimed book, *Sustainable Energy – Without the Hot Air*. He had also been Chief Scientific Advisor to the UK government's Department of Energy and Climate Change. Tragically, MacKay lost his battle against stomach cancer on 14th April 2016. He was 48 years old.

During Dowling's tenure, the Department's contributions to teaching and research were more widely reported than ever before, and its ability to attract major funding from industry, such as the partnership with Rolls-Royce, was brought to the attention of other University departments and faculties. In her role as Head of Department, Dowling was able to champion gender and ethnic equality and diversity not just within the Department but also throughout the University.

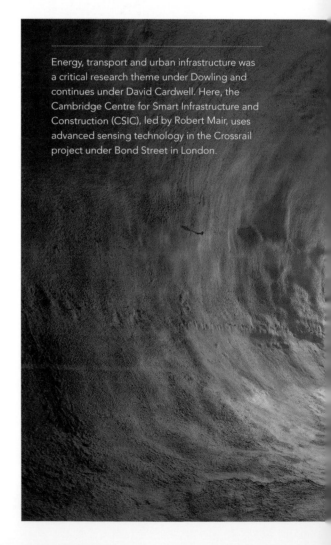

Energy, transport and urban infrastructure was a critical research theme under Dowling and continues under David Cardwell. Here, the Cambridge Centre for Smart Infrastructure and Construction (CSIC), led by Robert Mair, uses advanced sensing technology in the Crossrail project under Bond Street in London.

RESEARCH UNDER DOWLING

Research in the Department during Dowling's five years is best described by summarising the data provided by the Department for the Research Excellence Framework (REF) in 2014. The Department's research was divided into six research groups, together with the name of the group leader and the number of academics participating in each research activity. In each group, there was also a large number of research students working under the supervision of the academic staff. The Department had grown considerably during Glover's tenure, and continued to grow quickly during Dowling's headship.

Overarching across the groups were four strategic research themes which brought together multidisciplinary teams comprising academic staff drawn from several different research groups: energy, transport and urban infrastructure; uncertainty, risk and resilience; bioengineering; and inspiring research through industrial collaboration. These individuals had the freedom to join a strategic research team either as their full commitment to research or in addition to research within their specialism. The themes were designed to address major challenges, generate an impact upon society, look towards blue-skies research and seek links in other University departments, as well as in universities across the world and in industry.

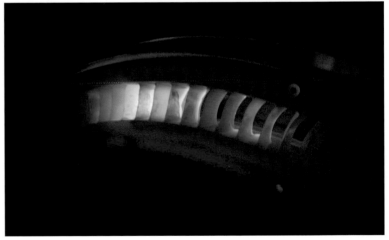

Above: Looking for flow irregularities on a set of turbine blades with brightly-coloured paint and light oil. This research was undertaken in collaboration with Rolls-Royce with the aim of improving the quality of flow through the blade row to improve fuel efficiency.

Top: David Cebon performs road freight research as part of the energy, transport and urban infrastructure strategic theme, which he also leads. His research is closely coupled to industry through his Centre for Sustainable Road Freight and his Cambridge Vehicle Dynamics Consortium.

SUCCESS IN THE RESEARCH EXCELLENCE FRAMEWORK 2014

The Department competed in the General Engineering unit of assessment. Whereas many of its rivals at other UK institutions handpicked its star performers, the Department submitted all but a few of its principal investigators – nearly 180 full-time equivalents. This huge submission demanded the production of 19 impact case studies presenting examples of how the Department's work had changed the world in forensic detail – the Department picked the best from a shortlist of over 40. In the overview statement about the Department, it was nearly impossible to convey its achievements within the strictly enforced page limit. Highlights included:

- one third of research grant income coming from industry, commerce and public corporations, which was a new high
- over 160 intellectual property disclosures leading to over 60 patent applications
- seven major spin-out companies that went on to raise over £15 million of external funding
- over 90 companies engaged with the Department in actively managed research relationships
- 164 honours, prizes and awards for the Department's researchers.

The results of REF 2014 of UK research showed that Cambridge had the greatest concentration of world-leading engineering research in the country (based on multiplying the proportion of research judged to be world-leading by the number of full-time equivalent people submitted). Ninety-five per cent of its work was considered either world-leading or internationally excellent in terms of originality, significance and rigour. The Department was also judged the best environment for engineering research with a perfect score unrivalled by any other General Engineering submission.

The bioengineering strategic research theme is showcased by the research of Graham Treece in this computerised tomography scan showing cortical thickness on a human femur and hip joint. Treece models the physics of scanners and applies novel processing methods to glean remarkable high-resolution data from unmodified conventional medical imaging instruments.

The Department continued to subject its strategy to review by the International Visiting Committee of leading national and international engineers, established by Glover and chaired by Broers. Following its deliberations, the Visiting Committee reported that 'The Committee was hugely impressed with the performance of the Department; by the way it is now acting as a whole rather than a series of separate groups in its strategic thinking and planning, by its connections to other departments and its coordination of many major bids to the benefit of the wider university.' This independent judgement gave the Head of the Engineering Department and all other concerned parties a great deal of reassurance and satisfaction.

In 2009/10, the Department took another novel step in redefining 'engineering' by placing sustainability at the very centre of its thematic description, asserting that engineering knowledge must foster sustainability, prosperity and resilience. It offered to share the knowledge it was acquiring with industry by every conceivable means, ranging from publication in the open literature to University-led entrepreneurship.

Malcolm Smith holds his inerter device, which was first deployed in Formula 1 racing and is poised to reach sectors beyond motorsports. Paddy Lowe, Executive Director (Technical) of Mercedes Formula 1 team since 2013 and alumnus of the Department, said: 'The Inerter ... has become a standard element of F1 suspension systems, now of equal rank to the spring and the damper in our constant search for higher levels of grip and stability.'

LAING O'ROURKE CENTRE FOR CONSTRUCTION ENGINEERING AND TECHNOLOGY

The Laing O'Rourke Centre for Construction Engineering and Technology is a partnership between the Department of Engineering, Cambridge Judge Business School and Laing O'Rourke, the UK's largest privately owned construction company.

The Centre was created as the result of a generous donation by Laing O'Rourke. The donation has funded a new Chair in Construction Engineering (Cam Middleton was elected in 2011), a Master's degree programme and supports research and undergraduate teaching within the Department of Engineering.

Ray O'Rourke said in 2010: 'I believe the engineering and construction profession has reached a critical crossroads in its development – it has for too long relied on traditional skills and approaches, often failing to keep pace with the political, social and economic demands of modern society. It must once again attract the very best talent, apply radical thinking, embrace new technologies and innovate in a way that removes waste and inefficiency and creates the greatest value for the world's communities. The goal of the Centre is to precipitate innovation and provide a new vision for the shape of tomorrow's construction industry.'

The Laing O'Rourke Centre moved to the James Dyson Building in 2016.

Among the highlights of the Department's plans was the declared intention to 'complete the Laboratory for Engineering for Life Sciences and to build stronger links between this theme and related activities in the Clinical School, and the Biological Sciences'. Later, the theme was renamed 'Bioengineering', and engineering was strongly linked to medical research. In 2010, the Laing O'Rourke Centre for Construction Engineering and Technology was founded, creating a new Master's and research programme to catalyse a revolution in construction engineering (see box).

Earlier mention has been made of Dowling's many honours and achievements. In 2016, she was appointed to the country's highest honour, the Order of Merit (OM), conferred upon very few individuals – just 24 hold the honour at any one time. Today, far from resting on these laurels, Dowling continues her research in the Department and is an ambassador for engineering worldwide. She brings public, government and corporate attention to the value of engineering, the demand for more engineers and the need for gender diversity in the profession.

In this story of the Department, we have journeyed from the pre-Victorian era to the current day: from a time when engineering was viewed with disdain by many academics to one in which it is seen as a jewel in the University's crown; from a handful of staff to the largest department; and from a driver for industry to a driver for global change. These changes will no doubt continue. The very latest can be found on the Department's website (www.eng.cam.ac.uk), where new tales of transformation are told to a worldwide audience every week.

Ray O'Rourke, founder of the Laing O'Rourke Group. His vision and generous donation enabled the foundation of the Laing O'Rourke Centre for Construction Engineering and Technology.

CHAPTER EIGHT

Teaching Matters

Picture the scene at nine o'clock on a Monday morning. From every direction, students arrive at the Engineering Department at Trumpington Street, on foot, by bicycle, singly, in pairs, in gaggles. The lecture theatres fill up – only the front rows are empty, while the lecturer paces up and down and riffles through notes. Technicians in laboratories all over the Department gather groups of students and settle them down. The drawing office, now known as the Design and Project Office (DPO), hums with activity as pairs of students perch on the tall stools and start work.

A graduate of the Department from 1935 or 1975 would find these scenes familiar, but some of the detail has changed dramatically. At least a quarter of the students are international, many of them Chinese; and at least a quarter are women. The DPO is now full of computers (although there are still some drawing boards too – probably the same ones). Students sitting in a lecture room may still be taking notes by hand, but some will be using tablet computers instead.

Allan McRobie introduces engineering to prospective students at the Department's Open Days.

THE ENGINEERING TRIPOSES

One thing that has emphatically not changed is the emphasis on a broad grounding in the fundamentals of all the main disciplines important to professional engineers. Graduates from past years, and potential employers of future graduates, regularly comment on how much they value this aspect of the Cambridge course. The Part I course, for the first two years, works students very hard and gives them very little choice. The course remains uncompromisingly rigorous: mathematical and analytical concepts are regarded as the foundations on which knowledge and understanding of engineering are developed.

Top left: A team of Manufacturing Engineering Tripos (MET) students with their exercise prosthetic for amputees, which the team developed during their final-year design project.

Top right: Teddy bears launched into space by the student-led Cambridge University Spaceflight team.

Above: David Cole lecturing to students on mechanical engineering.

Right: Tore Butlin demonstrating principles of angular momentum.

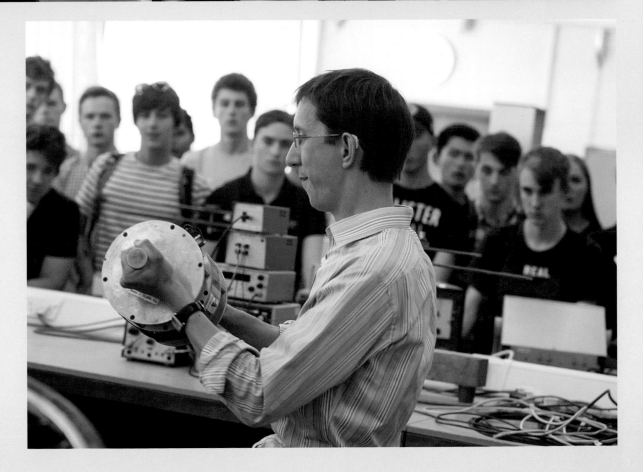

A DAY IN THE LIFE OF A SECOND-YEAR STUDENT

Out on the river at seven, just back in time for the nine o'clock lecture. Control theory: rather mathematical, but enlivened by the lecturer demonstrating a LEGO robot. Then Mechanics: today gyroscopes, demonstrated by the lecturer throwing a boomerang round the room over our heads. A bit mind-bending: didn't quite understand the explanation, but able to check his online material and videos afterwards and it seems clearer now.

Grab a cake and a drink, then into a lab session: Integrated Design Project today, programming our team's robot and struggling with a problem with the sensors.

Raining, so ate lunch in the CUED canteen rather than riding back to college. Then into the Dyson Centre to set up 3D printing of a component for the IDP robot. While that was printing, worked on problems in the library in last-minute attempt to finish an examples sheet before a five o'clock supervision.

Back to Newnham for the supervision, then out to a rehearsal of the University Chinese Orchestra.

Experimental work is integral to the course, with labs, computer exercises and project work used to illustrate concepts, build familiarity and instil a healthy scepticism about the infallibility of 'theory'.

In Part II, now the final two years of a four-year course, students have a wide choice among specialised modules covering all the main engineering disciplines – and, of course, the spread of these disciplines is significantly wider than in earlier years. Mechanics, structures and electrical machines are still taught, but so are biomedical engineering, signal processing, machine learning and robotics, among many other subjects. There are courses on management topics, required by the engineering institutions and now mostly taught by staff from the Judge Business School, which itself originated as a spin-out from the Engineering Department. There are language courses run by the Department's own language unit, which makes resources available to all undergraduates and graduates.

Alongside these lecture courses, there are projects in each year, increasing in scope and complexity through the course. All undergraduates remember their first year Structural Design project in which they design, build and test to destruction a cantilever bridge. In the final year, Part IIB, every student does an individual year-long major project. The variety of topics and nature of projects is vast: they may include

Above: Daniel Strange and Michelle Oyen using a robot constructed from LEGO to make synthetic bone.

Above right: Three-dimensional printer available for use by students in the Dyson Centre for Engineering Design.

MyMax, a Manufacturing Engineering Tripos (MET) final-year student design project, is an optical device that can be clipped to the front of a standard projector to show films in 3D.

development projects with industry involvement, or they have the character of research projects, and indeed may feed into other Department research activity and lead to publications in the scientific literature. Others have a 'design and build' flavour, some being linked to student-led group projects like the Eco Racing team (see box on page 140), and others aimed at engineering solutions to assist developing countries.

The combined effect of all this is that a Cambridge engineering graduate can assemble a state-of-the-art education in one (or indeed more than one) of the major professional specialisations. The course is accredited by an impressive range of professional bodies: in addition to the traditional 'big three' Institutions of Civil, Mechanical and Electrical Engineers (the last now re-branded as the IET), there is the Royal Aeronautical Society, the Institution of Structural Engineers, the Institute of Highway Engineers, the Institution of Highways and Transportation, the Institute of Measurement and Control, and the Institute of Physics and Engineering in Medicine.

The course organisation acknowledges this diversity by offering nine designated 'Engineering Areas', each specifying a range of choice from among the third- and fourth-year modules. These ranges overlap, so that most students graduate with a formal qualification in more than one area: for example, a student may qualify in both Mechanical Engineering and Bioengineering.

One of the areas of specialisation is Manufacturing Engineering: unlike other areas this is a separate Tripos (MET), running in parallel with Engineering Part II, with competitive entry for a limited number of students (around 40). As

Research student Sakthy Selvakumaran (right) stands with a group of fourth-year undergraduate students during a visit to the Crossrail Liverpool Street station.

COURSE STRUCTURE SUMMARY

PART I ENGINEERING (TWO YEARS)

General engineering for first five terms, no choice. Sixth term: elective lecture courses.

Lecture courses

Mechanics, structures, materials, fluid dynamics, thermodynamics, electrical engineering, electronics, information, control, mathematics.

Experimental work

Approximately two 2-hour laboratory experiments per week covering all subjects, plus: engineering drawing exercises; computing projects: structural design; integrated electrical project; integrated design project 'Robot lab'; integrated coursework 'Buildings in earthquakes'; mini-project 'Materials characterisation'.

Assessment

Part IA: Four 3-hour examination papers, no choice of questions. 10% credit from experimental work.

Part IB: Mainly 2-hour examination papers, choice (broadly) four questions out of six. 15% credit from experimental work.

After Part I:

Specialise into engineering area of choice, or keep a broad focus.

Engineering areas:

Aerospace and aerothermal engineering; mechanical engineering; bioengineering; civil, structural and environmental engineering; energy, sustainability and the environment; electrical and electronic engineering; electrical and information science; information and computer engineering; instrumentation and control.

PART II ENGINEERING (TWO YEARS)

Part IIA

| Choice of 10 modules out of 45. | 16 hours of lectures per module. | Labs/coursework associated with each module. Two projects in Easter Term. | 1.5-hour examination paper for each module. 30% credit from experimental work. |

Part IIB

| Choice of 8 modules out of 75. | 12–14 hours of lectures per module. | Modules may include coursework. Individual project runs through the year. | 1.5-hour examination for some modules; others are assessed by coursework. 43% credit from project. |

Graduate with Honours degrees of Bachelor of Arts (BA) and Master of Engineering (MEng)

Test equipment being used to monitor temperature and humidity at a prototype house at Jesus College. The project is part of the EcoHouse Initiative, a student-led group with the goal of improving sustainability and living standards in Latin America.

described in Chapter 6, the course was set up in 1979 as the Production Engineering Tripos (PET), to fast-track high-calibre engineering graduates into manufacturing industry, and its unique structure in the final year combines academic courses with periods out in industry putting theory into practice by solving real problems. Students, working in pairs, are typically away from Cambridge for the two- or four-week projects, staying in bed-and-breakfast accommodation and working in the company alongside employees. They have an industrial supervisor within the company providing some day-to-day support, but they are very reliant on their own knowledge and resources. Their Cambridge supervisor visits a couple of times during the project, but is otherwise only available on the end of a phone or by email, so students must become very self-sufficient. In the course of the projects, students make presentations to companies, to audiences that may include the managing director as well as shop-floor workers: it all contributes to a powerful learning experience. MET graduates are archetypically confident and articulate, and are much sought-after as employees because they can quickly start being directly useful to the company: the course is highly rated by manufacturing industry and by the engineering institutions. The small cohort size, the emphasis on teamwork and the challenges of the project work away from Cambridge encourage the formation of closely-bonded groups, contributing to the popularity of the course and the loyalty of its alumni.

Above: Hugh Hunt demonstrates the engineering principles of a boomerang to visiting schoolchildren.

Below: The Cambridge Autonomous Underwater Vehicle is an undergraduate project to design and build robotic submarines that can complete missions without human intervention. The team here is taking the vehicle apart after a day of testing in the water.

EXTRACURRICULAR PROJECTS

Undergraduates have often been pleased to have friends in the Engineering Department to help with important challenges such as fixing a bike or building a widget. The souped-up beds built for the charity fundraising RAG bed races of the 1960s and 1970s were made in odd corners of the Department, relying on goodwill from academic and technical staff. Bed frames were equipped with bicycle wheels and some form of steering mechanism, and taken

CAMBRIDGE UNIVERSITY ECO RACING (CUER)

Founded in 2007, CUER designs, builds and races solar-powered vehicles.

Evolution travelled 2,047 km on solar power across the Australian Outback as part of the 2015 World Solar Challenge. Powered by a combination of static and deployable photovoltaic panels and lithium ion cells, the car can travel at speeds of up to 110 kph. Innovative features of *Evolution* have been researched and developed by a team of around 60 students, with input from staff in the Engineering Department, and have focused on the aerodynamic design, the solar array, software for control and manufacturing processes aimed at reducing vehicle weight.

Above: The student-led Cambridge University Eco Racing (CUER) team tests their solar car for stability.

Below: The silent study area of the Engineering Library in 2016.

out on public highways. In her radio interview on *Desert Island Discs* (July 2016), Dowling recalled first meeting her future husband Tom Hynes whilst piloting a bed that he was pushing in a bed race from Cambridge to London. The recent opening of the Dyson Centre (funded substantially by the James Dyson Foundation) and the Oatley Garages (enabled with funding from the Oatley family) has at last provided facilities and resources to allow students to work on their own engineering projects. The Dyson Centre was created in space liberated by stripping out the Upper Workshops and offices on the first floor of the Inglis Building, and is used for project work associated with the formal course as well as for extracurricular enterprise. It has a full-time manager and technician support, and is equipped with tools for prototyping and making – from traditional hand- and machine-tools to computer-controlled machinery, 3D printers and laser cutters. Students are encouraged to do more design, creation and innovation work, and to try out their ideas in the Centre.

Serious project work requires funds, and the Student-led Project and Industry Partnership (SPIP) provides this by inviting industrial partners to provide financial, technical, managerial and mentoring support. Projects supported range from Cambridge University Eco Racing (see box) and Full Blue Racing, both with teams of around 60 students, through to the three-student team Project Voxel, developing innovative software and electronics for novel lighting effects.

LIBRARY

Libraries throughout the University are re-inventing themselves for the online generation, and the Engineering Department Library is leading the way with its newly restructured space. There are still books and journals, but

The student-led Full Blue Racing team work on their car in the Oatley Garage space.

for many the internet is the main source of information. Independent study spaces full of students working on examples papers and coursework in a spare hour between lectures will be familiar to many generations of students. But there are now also groups working together in the collaborative space, with jigsaws, colouring books and bean bags to promote contemplation and diversion. Library staff provide information and guidance on a great range of topics relevant to students, including writing reports and essays, searching and referencing the literature and dealing with copyright. The Library is at the heart of the Department, providing a versatile study centre to suit all tastes and requirements.

TECHNOLOGY OF TEACHING

Lectures remain the main route for the transmission of information, but the surrounding technology has evolved with time. Few now favour chalk and blackboards, although whiteboards and coloured pens are still on the agenda in smaller rooms. Overhead projectors now feel old-fashioned, with 'visualisers' taking their place. Computer projectors, two in every lecture room, are commonly used to display the standard partially-completed notes that students fill in during lectures. Most lecturers provide paper copies of handouts, but the trend is towards purely electronic resources, with students taking notes using tablets or computers in lectures. Manufacturing Engineering has taken this approach and has been paperless for some years now; the feasibility of extending this approach to other courses in the Department is being explored.

OUTREACH AND ADMISSIONS

Visitors to the Department are sometimes taken by surprise to hear loud and enthusiastic young voices: 'Five, four, three, two, one ... WHOOSH!' These are groups of primary school pupils who have built compressed-air powered 'rockets' under the guidance of the Department's Outreach Officer, Maria Kettle, and her crew of student volunteers, and are launching them in the central courtyard, hoping to clear the roof of the Centre Wing building.

The Department works hard to interest young people in engineering and science through its outreach programme and open days. The Outreach Officer, student volunteers and many academic staff are involved, with several academic staff members having won awards for their outreach activities. Indeed, Hugh Hunt has become something of a media star, reaching huge audiences by fronting several technology-based TV programmes (see Chapter 9). As well as engaging with primary schools, the Department runs events and activities for secondary schools and families, in addition to some aimed specifically at prospective applicants, particularly from the state sector. This

UNDERGRADUATE ENGINEERING ENTRY STATISTICS, 2015

Total: 309 students; 228 male (74%); 81 female (26%).

Nationality: 202 UK; 26 China; 14 Malaysia; 8 Singapore; 7 India; Cyprus 6; 46 from 30 other countries.

School type: 42% maintained; 32% independent; 26% overseas.

Intended specialisation: 14% mechanical; 11% aeronautical; 11% information; 10% chemical; 7% civil; 8% electrical; 3% biomedical; 3% general; 2% manufacturing; 31% other.

Interests: 43% design; 33% working abroad; 27% development; 22% entrepreneur; 19% research; 19% sustainability; 17% multidisciplinary; 11% management; 5% finance; 3% teaching.

Schoolchildren attend a Cambridge Robogals chapter outreach event at the Department in 2014. Robogals is a global non-profit outreach group dedicated to introducing young girls to science and engineering careers through robotics workshops and competitions.

may involve travelling to schools with organised masterclass sessions, but often the most productive events involve getting schools to come to Cambridge so that advantage can be taken of local facilities, from LEGO robot kits to the 'Make-Space' in the Dyson Centre. Some of these activities are integrated with the University-wide Science Festival, held annually during the Easter vacation.

Robogals is an international initiative aimed at increasing the number of women taking STEM subjects at school, and perhaps inspiring them to go into engineering. A number of half-day Robogals sessions have now been run in the Department, where girls compete in groups to get robots to perform a task.

Engineering applications to Cambridge are buoyant, bucking the trend for many other UK universities. Some statistics for the 2015 entrants are given here (see box). The course is thoroughly multicultural, and a contributory factor to the strength of the international applicants is that an admissions team of Cambridge staff is sent out to China, Malaysia and Singapore, interviewing the highest calibre applicants in person.

RESPONDING TO CHANGE

Students entering engineering now have different skills and experience from what would have been seen 50 years ago, or even 10 years ago. School education in the UK has been through times of change, with reduced emphasis on the mathematical and analytical content of A-level courses. In addition, lifestyles have changed. The make-and-mend mindset of earlier times is now the exception rather than the rule; tinkering with cars and radios is no longer productive; fitting an electric plug is vanishingly rare. Many students arrive at Cambridge having never handled a screwdriver or spanner, and with no intuitive understanding of mechanical devices. The engineering course has had to adapt in style and content to help students to bridge the gap between school and university. A very successful venture has been to start the course by dividing the students into teams of three, doing a design-and-build project using LEGO Technic over a 10–day period. They construct working devices, sometimes very impressive: examples have included a photocopier, an alarm clock and a coin sorter.

THE FUTURE

With the prospect of a move to West Cambridge and designed-for-purpose buildings, the Department is having to guess what university engineering education will look like in the future. Will lecturers still be standing in front of a class of 320 students in the same room – indeed, will they be delivering lectures at all? A University pilot scheme for recording lectures, 'lecture capture', is being trialled in the Engineering Department; there will be careful scrutiny of the way in which students use lectures that are available on the web. Assessment by hand-written examinations is already feeling anachronistic: but is there a workable electronic alternative? Will the Cambridge supervision system survive, or will financial pressure force change? Different learning models proliferate in reputable institutions over the world, and Cambridge is looking around to see what might work for us: radical ideas include problem-based learning, so-called 'flipped classroom' teaching, and MOOCs (Massive Open Online Courses). The encouraging news is that the Department's teaching staff, especially the younger ones, are showing great enthusiasm to explore these new possibilities. They really believe that 'teaching matters', and they will ensure that the Department remains a world-leading teaching institution.

EMPLOYMENT: PRINCIPAL DESTINATIONS OF ENGINEERING GRADUATES, 2015

16% further study; 17% IT sector; 17% manufacturing; 11% engineering consultancy; 6% management consultancy; 2% banking; 31% other.

Event led by Outreach Officer Maria Kettle to test cantilever bridges built by schoolchildren.

CHAPTER NINE

Changing the World

Engineering is nothing if it does not change the world. This chapter captures a handful of the Department's successes over recent decades – in terms of both development within the Department and spin outs from the Department. Meanwhile, the Department's graduates, publications and media work take our teaching and research to a vast global audience every day. Online resources, free open access to published papers and social media are accelerating the pace and extending the reach. The latest news on the Department can be found on its website (www.eng.cam.ac.uk), where links to its news stream and media channels can be easily found.

The West Cambridge Diaspora

The Whittle Laboratory, the Schofield Centre, the Institute for Manufacturing, the Nanoscience Centre and Electrical Engineering: these five integral parts of the main Engineering Department are already well established in West Cambridge. The history and activities of these research centres begins this chapter.

THE WHITTLE LABORATORY
The Whittle Laboratory is internationally recognised as a centre of excellence for research in turbomachinery aerodynamics. Over its 43 years of history, the Whittle has worked with leading key industrial partners around the world, including Rolls-Royce, Mitsubishi Heavy Industries, Siemens and Dyson. The academic excellence of the Whittle is evidenced by more than

A false-coloured high magnification electron micrograph of free-standing graphene foam.

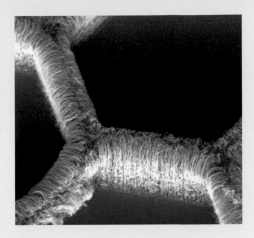

Above: Carbon nanotubes grown in a honeycomb structure using electron beam lithography.

Left: Centre for Smart Infrastructure and Construction (CSIC) engineers installing fibre optic cables in a tunnel below London.

Below: Researchers in the Robotics Laboratory creating new robotic designs.

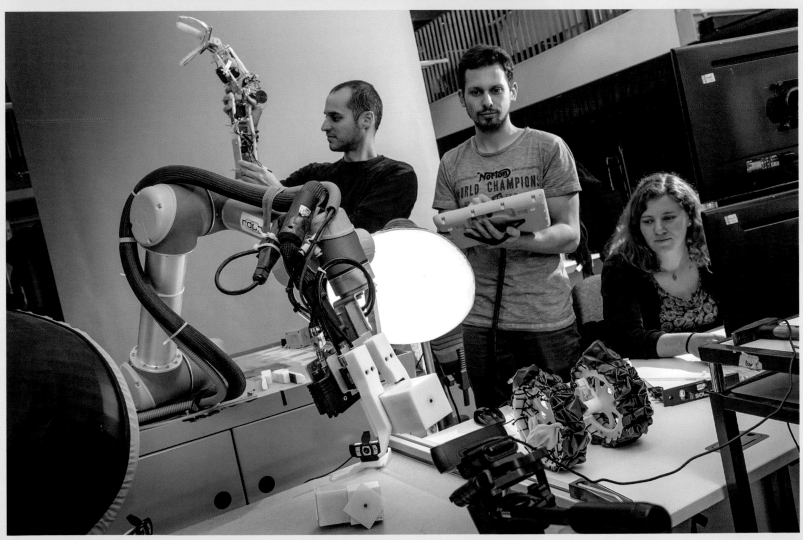

30 American Society of Mechanical Engineers International Gas Turbine Institute Annual Best Paper awards – more than any other single institution – as well as numerous prizes from the American Institute of Aeronautics and Astronautics, the Institution of Mechanical Engineers, the Society of Automotive Engineers and others. The Whittle has also been active and successful in researching, publicising and commercialising computational fluid dynamics (CFD) software to simulate turbomachinery aerodynamics. This software has been licensed around the world to over 50 companies and research institutes, and is the reference standard today. Several associated commercial companies have been spun out of the Whittle, to develop and service descendants of this software.

Throughout its history in the Engineering Department on Trumpington Street, research in turbomachinery aerodynamics was hampered by a combination of lack of space and the unacceptable noise levels generated during investigations of high-speed flows in the turbomachinery. So uncomfortable and intrusive was the noise that tests were virtually prohibited. In 1969, William Hawthorne and John Horlock decided, in partnership, that more space and separation from the Scroope House Estate was essential

Above: Light scattered from micrometre-sized olive oil particles carried by a fuel–air mixture as it enters a combustor during a blow-off event.

Below: A researcher in the Whittle Laboratory examines a turbine test rig.

Frank Whittle at the opening of the SRC Turbomachinery Laboratory in 1971.

The SRC Turbomachinery Laboratory, later renamed the Whittle Laboratory.

for their ambitious plans for a world-class turbomachinery laboratory in Cambridge. Fortuitously, Horlock was Chair of the Mechanical Engineering Committee of the Science Research Council (SRC), and he had observed with interest, and perhaps with some envy, that science departments in universities were able to secure money from the Council for buildings to house their equipment, whereas the Engineering Department had not applied to the SRC for a building in its entire history. In agreement with Hawthorne, he felt that the time was ripe for Engineering to make a bid for a building, particularly as Hawthorne knew that the University owned suitable land in West Cambridge, two miles from the main site, which could be made available at no cost.

Horlock's proposal to the SRC for a major turbomachinery research initiative in support of British industry was successful. A sum of £272,000 was awarded for the construction of a turbomachinery laboratory, and the first University research laboratory, in what is now a burgeoning high-tech city in West Cambridge, was built. Work commenced in December 1969 and the SRC Turbomachinery Laboratory was available for occupation in May 1971. Horlock was appointed Founding Director of the Laboratory and Paul Gostelow the Deputy Director. Gostelow wrote: 'With compressible flow work becoming basic to the needs of industry a move has been made to the new site which can accommodate suitable new high-speed equipment together with the large number of existing low-speed test facilities.' On 25th May 1973, the Laboratory was opened formally by Air Commodore Sir Frank Whittle and renamed the Whittle Laboratory in recognition of his invention of the jet engine and his association as an undergraduate with the Engineering Department.

Until 1974, Horlock and Gostelow were the only full-time academics working in the new Laboratory, but they were well supported by research students, postdoctoral staff and technicians. Some academics, notably Denis Whitehead, Michael Cooper and Homayoon Daneshyar, spent roughly one day per week at the Whittle Laboratory. Nick Cumpsty joined the Laboratory in 1974 and John Young, with a particular interest in wet steam, was recruited at the same time. In 1974, there was an upheaval when Horlock unexpectedly resigned and Hawthorne had to step in as Director, adopting a hands-off approach, but SRC support now waned – it was felt that the leadership shown by Horlock would be missing. Fortunately, Rolls-Royce stepped in to support the Laboratory through a link between Ffowcs Williams and Stanley Hooker, Technical Director of Rolls-Royce, and the collaboration has remained strong ever since.

Nick Cumpsty, Emeritus Professor of Aerothermal Technology and Director of the Whittle Laboratory from 1990 to 1998.

Bill Dawes, current Director of the Whittle Laboratory and Francis Mond Professor of Aeronautical Engineering.

When Hawthorne retired, Whitehead took over as Director, again in a hands-off style, but matters improved when John Denton was recruited in 1976 as a full-time lecturer, bringing links to the steam turbine industry and the Central Electricity Generating Board.

In 1984, Denton became Director of the Whittle Laboratory, serving in that position until 1990, when he was succeeded by Cumpsty. When Cumpsty left Cambridge, Denton served a second term as Director until his retirement in 2005. In 1984, Bill Dawes, having finished his PhD in Cambridge five years earlier, was recruited back to the Whittle, and there was now a real core focus on computational fluid dynamics (CFD) in the Laboratory, which was able to purchase fast computers with proceeds from the sale of its software. In 1996, Bill Dawes was appointed Francis Mond Professor, succeeding Michael Gaster.

The core philosophy of the Whittle is based on combining experimental measurements with computer simulations to develop and exploit a physical understanding, working in close collaboration with industry. The Whittle vision is to extend a research arc spanning innovative fundamental studies all the way to successful applications of new ideas in production engines. With a core of bright young lecturers, a series of benign but engaged directors, new staff and the continual throughput of highly capable research students, research in experimental and modelling work has advanced at great pace. The ethos of hard work and high standards was and is accepted by all staff, who gather round a single table for morning and afternoon tea breaks – a vital mingling from junior technician to senior professor, all committed to high research standards. The Whittle was always internationally linked and outward looking, and almost every day visitors from around the world are welcomed to meet at the 'Tea Table'. Regular seminars have always been used to disseminate research results, and the Laboratory was soon recognised, just as it is today, as the premier place for the study of turbomachinery, for the quality of its original research and its enterprise in spinning out successful commercial companies.

Coinciding with the early years of the Whittle Laboratory, in the 1980s the University established the Wolfson Cambridge Industrial Unit, led by Donald Welbourn to improve and encourage interaction with industry. Part of the Unit was Lynxvale Ltd, led by Stephen Bragg, former Vice-Chancellor of Brunel University, which acted as a commercial vehicle for licensing University software, technology and expertise. (Today the Unit has

transformed itself into Cambridge Enterprise.) Denton and then Dawes were quick to embrace the new world of commercialising University research. They joked that the new conservatory built behind Lynxvale's tiny offices in 20 Trumpington Street was paid for with royalties from licensing their software. The 'Denton Code' and, independently, the 'Dawes Code' were taken up for use by commercial and research organisations across the world, and these software packages became industry standards, changing the world of turbomachinery research and development.

Turbostream, the evolution of the Denton Code developed by Denton in the 1970s, is now used extensively for the design of new commercial machines by a commercial company, Turbostream Ltd, and by the Engineering Department's Whittle Laboratory for academic research. In parallel, the descendants of the Dawes Code are being exploited commercially by Cambridge Flow Solutions Ltd, and are also used within the Whittle for CFD research.

The Whittle is widely recognised today for its successful activity in fluid dynamics research, for writing code and for commercialising CFD software to simulate turbomachinery aerodynamics.

Rob Miller (right), Professor of Aerothermal Technology, explains the use of the novel high-temperature probe developed by him and his team.

Right: Andrew Schofield, Emeritus Professor of Geotechnical Engineering, after whom the Schofield Centre for Geotechnical Process and Construction Modelling is named.

THE SCHOFIELD CENTRE

At a ceremony on 12th September 1998, the Vice-Chancellor, Alec Broers, named the geotechnical centrifuge laboratory in West Cambridge the Schofield Centre in honour of Andrew Schofield, the man who had devoted his entire working life to reaching a clear understanding of the principles of soil mechanics. By devising testing methods that were then followed by experimentation and validation of the results, he had brought a deeper understanding to modern geotechnics and swept away out-of-date ideas and beliefs. Following the formal naming ceremony, there was an Open Day at the Centre and a retirement dinner for Schofield, at which there were tributes recognising his extraordinary research contributions and inspiring leadership over two-and-a-half decades.

The history of the Soil Mechanics Group began with John Baker accepting Ken Roscoe as a research student in 1945, just months after his escape from a German prison and return to Cambridge, where he had been an undergraduate. Roscoe worked on the strength of soils, devising a simple shear apparatus (SSA) for experimental research and, with Baker's encouragement, building up a soil mechanics laboratory. Progress was slow until 1954, when he started a study of lateral earth pressure against a rotating short pier foundation and a study of soil in the SSA, working closely with his

early research students. As more research students joined, Roscoe embarked upon soil radiography with the help of Addenbrooke's Hospital, and used scanning electron microscopy to study soils. The research gained momentum when Roscoe was promoted to Reader, as he described in the 10th Rankine Lecture, 'The Influence of Strains in Soil Mechanics'. With the availability of new computational facilities, the group was able to explain the observed behaviour of soils as an elasto-plastic material, postulating the 'Critical State Theory', which changed the perspective of the world's geotechnical community. Roscoe died in a car accident in 1970, just two years after his promotion to a Chair; Peter Wroth then took up the leadership of the research group until his long-standing colleague Andrew Schofield was appointed to the Chair that Roscoe had held.

Schofield was convinced that geotechnical centrifuge-based research on soils was essential. Following his appointment, he developed geotechnical centrifuge facilities believing that, with carefully designed experiments, geostatic stresses at great depths could be replicated. Today, the most prominent work in the Schofield Centre is the recreation and observation of failure mechanisms in geotechnical problems and their modelling before they occur in the field. In offshore wind farms, the long-term behaviour of monopile foundations that support wind turbines has been studied by replicating the wind loading on the foundations from the thousands of storms that would inevitably occur during a monopile's anticipated lifetime. The key finding was that the accumulated rotation of the monopiles after each storm would limit their operational life significantly in some of the soil stratigraphy found along Britain's coastline.

The centrifuge facility has also been used to study earthquake-induced soil liquefaction, whereby solid ground transforms under earthquake loading into a semi-liquid that flows like sludge. Buildings settle, rotate and collapse, leading to the devastated cityscapes seen after the Christchurch, Tohoku and Haiti earthquakes. Earthquake actuators have been developed at the Schofield Centre to recreate liquefaction in a soil body while it is spinning in the centrifuge, allowing the prediction of failure patterns in structures from buildings to bridges. Retrofit measures have been tested for their efficacy in mitigating the effects of liquefaction; for example, by injecting small quantities of air into the ground before an earthquake strikes, the settlements suffered by buildings can be reduced. Future research in the Schofield Centre will encompass novel and innovative processes for improving the resilience of the built environment to natural hazards.

Schofield was succeeded by his former student Robert Mair as Head of the Geotechnical Group and, at the turn of the millennium, the research group celebrated the announcement of a grant from the government for an extension of the Centre and the founding of a Centre for Geotechnical Process and Construction Modelling to address the key questions in construction and environmental issues. Research still focuses around the 10-metre-diameter geotechnical centrifuge designed and built in the Department under the supervision of Philip Turner and now named after him, the Turner Beam Centrifuge. (In his will, Turner endowed a prize, awarded annually, for outstanding work in geotechnical centrifuge testing by a research student.) New facilities have also been installed: earthquake simulators, multidegree robotic actuators and pioneering particle image velocimetry techniques for capturing accurate deformation behaviour in geotechnical models during high-G centrifuge tests, which have been adopted by almost 100 centrifuge research facilities worldwide.

The current Head of the Geotechnical and Environmental Group, Gopal Madabhushi, describes the mission of the Schofield Centre, where he works with a growing number of academic staff, graduate students and support staff, as follows: '[It focuses] on the key questions of construction and resilience of the built environment, where field, computational and physical modelling studies can be integrated in collaboration with industry.' Through its history, the group has produced more than 250 graduates with PhD degrees and continues to receive extensive support from the government and the civil engineering industry.

The 10-metre Turner Beam Centrifuge in the Schofield Centre, which is in constant use for long-term research and fast-turnaround work in emergencies. It is used to study topics as diverse as the resistance of buildings to earthquakes, the performance of monopile foundations in offshore wind farms, and tunnelling under buildings in cities.

THE INSTITUTE FOR MANUFACTURING

The Institute for Manufacturing (IfM), a division of the Department of Engineering, moved into its permanent home in November 2009 when the Chancellor of the University, HRH The Duke of Edinburgh, opened the striking Alan Reece Building in the presence of the principal donor and 400 well-wishers, bringing the Manufacturing Group's 50 years of nomadic existence to an end. Alan Reece's generous donation matched that of the Gatsby Charitable Foundation to raise £15 million, with support from Department funds and Marshalls, for the building in West Cambridge – an ideal site for the Institute's interactions with the scientific and technological departments that were moving steadily westwards. The Institute's expertise in industrial management, technology, policy, research and education, all designed to assist companies ranging from start-ups to multinationals to achieve their strategic goals, was on show. The Vice-Chancellor, speaking on the occasion, said: 'There is no more vivid a model of linking academia with the needs of society than the Institute for Manufacturing.' She added: 'In six-and-a-half years, I have been on an incredible journey with the IfM, which has seen it accomplish so much.' Under the leadership of Mike Gregory, the IfM has grown from being a basement activity in Mill Lane to one of the most prominent centres in the University of Cambridge.

Manufacturing's chequered history began in 1953 when Hawthorne, on returning from the USA with experience of modern industrial training, found the outmoded two-year apprenticeship programme for graduates still in existence and, recalling his own dreadful experience of apprenticeship before escaping to MIT, presented a paper to a conference of industry education officers entitled 'Professional Training for the Engineering Graduate'. In it he described the time-wasting, boredom and ridicule experienced by incredibly capable graduates who had to stand and watch factory workers for hours, always fearful of humiliation if they dared to offer the slightest comment on transparently outdated and time-consuming working practices. Instead of this 'trial by forced inactivity', Hawthorne proposed a programme of lectures,

Above: Alan Reece (left) and Mike Gregory at the ground-breaking ceremony for the Institute for Manufacturing's new home, which was named the Alan Reece Building to recognise both his generous gift that enabled the project to go forward and his inspirational lifelong contribution to engineering.

Below: The Manufacturing Engineering Tripos (MET) Design Show is a yearly highlight for the Institute for Manufacturing (IfM). Three innovative projects from 2016 were, from left, the BrekTech Wafflestation, a mechanical waffle maker; Paperbot, an automated papier-mâché machine; and Fruiticle, a cocktail ice product.

experiments, workshop training and reporting, and he accepted with alacrity the challenge from a training manager to devise a system. John Reddaway and David Marples were delegated by Hawthorne to prepare a course based on the material taught to undergraduates taking the ordinary degree, which included lecture courses on production methods. Two groups, each of 12 undergraduates, took part in short courses in 1954 and 1955 with conspicuous success, and the new concept was reported in a paper, 'An Approach to the Techniques of Graduate Training', but funds were unavailable for more courses and the scheme was abandoned.

John Baker, Head of the Engineering Department, supported the idea of a one-year course, and named it the 'Reddaway Plan' when it was presented to the Cambridge University Engineering Association's meeting in 1964. In the audience was Mike Sharman, and his enthusiasm was detected by Reddaway, who approached him to take charge of the course, for which funds were available for two years of operation. The Advanced Course in Production Methods and Management (ACPMM) was operating within a year under Sharman's enthusiastic leadership. The course emulated industrial disciplines and working practices in executing 'real projects', with factory visits and lectures by industry professionals and university academics. To the surprise of the industrial supporters, some projects were spectacularly successful, and all who took the course sang its praises. Among them, in the fourth year of the course's operation, was Mike Gregory, who later claimed that for him the experience was life changing. But the course was in constant jeopardy, occupying an anomalous position within the University, which occasionally tried to shut it down until, in 1984, it was taken over by Wolfson College. A design option was added and the course became the Advanced Course in Design, Manufacture and Management (ACDMM), but it still did not have standing in the University until Colin Andrew was appointed to the Chair of Manufacturing. He created the qualification of a diploma for the course and financial stability was established when the Gatsby Charitable Foundation provided five years of funding.

Following Sharman's retirement, the course became an MPhil in Industrial Systems Manufacture and Management (ISMM), with the customary dissertation at the end of the course. Its success was immediate, attracting five times as many applicants as there were available course places, and appealing to candidates with outstanding academic qualifications from across the world. At this stage, through government intervention, the Production Engineering Tripos (PET), later the Manufacturing Engineering Tripos, was established and Gregory, who had been on the staff of the ACPMM since 1975, took charge of it. When Gregory moved on, the course became the responsibility of Claire Barlow and continued to flourish, with its graduates going on to found and manage start-ups, transform existing manufacturing organisations by introducing

Andy Neely, head of the Institute for Manufacturing (IfM), in the Distributed Intelligence and Automation Laboratory (DIAL).

Distributed Intelligence and
Automation Laboratory (DIAL).

modern production and marketing methods, and design new
products and services.

Manufacturing's share of national productivity shrank in
the Thatcher decades, with its focus on services, while Germany
and Japan advanced spectacularly in the manufacture of
goods, leading the thrust towards globalisation of markets.
It was at this stage that Gregory proposed an Institute for
Manufacturing to Andrew, but it wasn't until 10 years later,
with his appointment to the Chair of Manufacturing and as
Head of a new Manufacturing and Management Division of the
Engineering Department, that Gregory's long journey came to
fruition when, supported by the Foundation of Manufacturing
and Industry, the IfM was formed as an independent entity
within the Engineering Department in 1998. During the
technological revolution at the millennium, new products and
companies were created at a rapid pace and 'disruptive' technologies eliminated
many established industries. Research, education and management practices were
also changing, and the IfM was awarded a major grant in 2001 and nominated
for EPSRC's flagship projects, growing to a significant size despite its cramped
premises in Mill Lane. The move to the Alan Reece Building at West Cambridge
relieved this constraint and gave the IfM its first purpose-built facilities.

All research at the IfM is based on the premise that it should be useful
to industry both in its subject matter and in its outputs. This was exemplified
by Ken Platts: recruited from industry to the University of Cambridge, he
published a workbook on behalf of the Department of Trade and Industry,
*Competitive Manufacturing: A Practical Approach to the Development of
Manufacturing Strategy*, which became a blueprint for industry and sold 10,000
copies, establishing Cambridge's credibility in manufacturing strategy, as did
Andy Neely in the area of performance measurement. Later, David Probert
created a framework for deciding whether a manufacturer should make
a product in its entirety or outsource parts to a supplier. His work created
robust technology management systems capable of helping companies turn
ideas into products and services. At the Centre for Technology Management,
Elizabeth Garnsey and her students conducted research into the emergence
from the University of high-tech firms and their growth paths, identifying

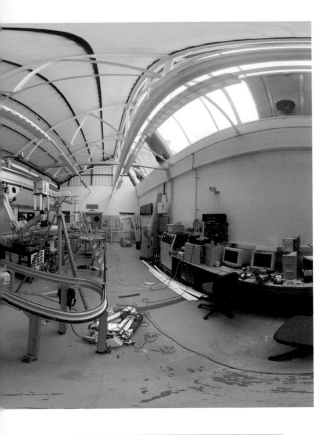

REDBITE SOLUTIONS LTD

Duncan McFarlane, together with former PhD student Alex Wong and lab manager Alan Thorne, went on to co-found RedBite Solutions Ltd, a company based in the St John's Innovation Centre. RedBite's vision is to identify, organise and enable intelligence for every single object in the world. Their products, Asset and Edge, provide asset and device management solutions, respectively. The unique software turns any object into a smart object that can participate in the Internet of Things. RedBite Asset involves providing a physical tag (QR code, RFID tag) or virtual tag (GPS) to objects. By scanning an object's tag using a smartphone, users can access the profile page of that asset via its unique URL. The profile page contains all historical asset data and maintenance records and allows users to contribute photos and comments and alert maintenance teams to any problems. RedBite Edge simplifies the process of connecting RFID and other Internet of Things devices directly to the cloud and allows for the remote management of a global network of connected devices. RedBite's software footprint currently spans 55 countries, with clients from a wide range of industries, and includes the United Nations, Bank of Tokyo, Manchester Airport and Costain.

Below: Duncan McFarlane, Head of DIAL, and Alex Wong, CEO of RedBite Solutions Ltd.

the special challenges facing those engaged in manufacturing. Yongjiang Shi contributed a new research area investigating international manufacturing networks which, under the leadership of Jag Srai, has since expanded to address the array of challenges faced by multinational companies when deciding on their 'manufacturing footprint' and supply chain configuration. More than 50 multinational companies have been helped to 'make the right things in the right places' and optimise their global supply chains. Supported by BAE Systems and the Royal Academy of Engineering, a new thrust into Engineering Services was introduced in 2003. Gregory, then Duncan McFarlane and most recently Neely led this area. He established the successful industry partnership Cambridge Service Alliance in 2010.

Another group, the Distributed Information and Automation Laboratory, led by McFarlane, worked on radio frequency identification for tracing and tracking components within the supply chain. McFarlane was also a co-founder of the Auto-ID Labs, a group of seven research centres that coined the much-used phrase 'the Internet of Things'. McFarlane and his team now study intelligent systems and smart data management for applications within factories and across supply chains. Asset management research is carried out in collaboration with the Cambridge Centre for Smart Infrastructure and Construction and Ian Hutchings, GKN Professor of Manufacturing Engineering, guides scientific research in inkjet printing and its applications. Bill O'Neill leads research in the Centre for Industrial Photonics on laser-based manufacturing processes and collaborates with Cranfield University in the EPSRC-funded Centre for Ultra-Precision Manufacturing Technologies. The availability of a large amount of laboratory space in the Alan Reece Building has enabled research teams to start work on carbon nanotubes and biosensors, with all scientific and technological projects pursued within the context of manufacturing.

Gregory believed firmly in the need to change government thinking on the importance of manufacture in a nation beguiled into believing that a service-oriented economy was paramount. For this reason, he introduced economics and policy direction research topics in the IfM, particularly posing the fundamental question, so often asked, of why some countries are so much better than the UK in translating ideas and research into products and economic advantages. Another initiative was the Manufacturing Professors' Forum, which brings together policymakers, academics and industrialists to discuss, on a broad base, the conditions needed for optimising manufacturing ventures.

As public and governmental consciousness of the need to preserve the planet and to maintain an environment acceptable to mankind increased year by year, the issue of sustainability in the acquisition of raw materials for production and in manufacturing processes became more and more important. The IfM is now heavily involved in sustainability research, with Steve Evans leading the Sustainable Manufacturing Group. The management of 'ideaSpace', the innovation facility in West Cambridge that provides flexible office space and networking opportunities for entrepreneurs and innovators, has been taken over by the IfM, and a thriving sector of start-up companies is supported through this arrangement.

Below and bottom: Examples of additive manufacturing performed at the Centre for Industrial Photonics.

Hawthorne's insight 50 years ago, that manufacturing teaching and research should be part of an engineer's education in industry and within the University, was the kernel that enabled the now-vibrant IfM, and manufacturing is now at the heart of engineering education. In its early days, the course was almost terminated from lack of funds, whereas now manufacturing receives more EPSRC funding within the University of Cambridge than any other university in the UK. The Manufacturing Engineering Tripos goes from strength to strength, and there are more than 75 students doing PhD or MPhil research in the IfM. For the future, the aim of the IfM is to remain committed to innovation and to take part in the development of a scale-up centre for the transition of ideas and concepts from prototypes into scalable industrial applications.

Mark Welland and members of his team at the Nanoscience Centre.

THE NANOSCIENCE CENTRE

Miniaturisation of the transistor, invented in 1947, was made possible by an understanding of the behaviour of materials at very small length scales and the development of tools for imaging and measuring, eventually down to a single atom. Ever more precise methods of fabricating transistors and circuits led to greater miniaturisation, and as research laboratories worldwide explored length scales smaller than one micron (one millionth of a metre) and reached the nanometre scale (one billionth of a metre) an important connection was made; arcing across the scientific disciplines of physics, biology, chemistry and materials science, the nanometre length scale opened up a completely new world of science and engineering. Whether it was the properties of engineered materials, the molecular processes that define life itself or the subtle physics of vanishingly small structures, the unifying theme was always the nanometre. The science and technology of the nanometre became known collectively as 'nanoscience' or 'nanotechnology', and the length scale was now just five atom diameters.

The opportunities for multidisciplinary science and technology research multiplied so rapidly that today there is hardly a university in the world without a nanoscience effort. In 1999, the Nobel Laureate Horst Stormer summed up the excitement of research in nanotechnology: 'Nanotechnology has given us the tools to play with the ultimate toy box of nature – atoms and molecules. Everything is made from it. The possibilities to create new things appear endless.'

Mark Welland, a Cambridge engineer, decided to explore the world of nanoscience at the start of his career. In the 1980s, Electrical Engineering at the

Engineering Department established the key elements of what would become nanotechnology through the research of Broers and Welland, who built the first scanning tunnelling microscope (used to image single atoms) in the UK. Their research groups explored both fabrication and measurement of a range of materials at the nanometre scale. When, in the mid-1990s, the UK government started to invest in nanotechnology, the research group was able to develop significantly by building a £12 million research facility in West Cambridge, the Nanoscience Centre, and securing a £10 million grant a few years later to establish an interdisciplinary research collaboration directed by Welland.

The Nanoscience Centre was a joint venture between the School of Technology and the School of Physical Sciences, led by Welland and the Cavendish Professor of Physics, Richard Friend, respectively. Their vision was to develop a research space that would make tools for fabrication and characterisation at the nanometre scale available to researchers from across the University of Cambridge and beyond. Welland moved all his research from the Scroope House site on Trumpington Street to the newly built Centre, an 1,800 m² research facility located at the eastern edge of the University's West Cambridge site. The Centre provides open access to clean rooms and low-noise laboratories to over 300 researchers, and is also the home of Electrical Engineering's Nanoscience Group.

A major research grant for an Interdisciplinary Research Centre in Nanotechnology established the Cambridge Nanoscience Centre as a world-leading interdisciplinary institution. Projects were funded across 15 departments in the universities of Cambridge and Bristol and University College, London, with the outcome of 30 interdisciplinary projects, 56 postdoctoral positions and 25 PhD studentships during the seven-year life of the original grant. More than 300 refereed articles were published, six patents granted, two commercial companies spun out and more funding attracted in order to establish nanoscience centres at Bristol University and University College, London.

In 2000, Russell Cowburn and Welland published a paper on room temperature magnetic quantum cellular automata in *Science*, pointing out

Above left: A cancer cell treated with gold and cisplatin, a conventional chemotherapy drug. These were released into tumour cells that had been taken from glioblastoma patients and grown in the lab.

Above: A spherical representation of carbon nanotubes grown from iron chloride and magnesium oxide.

Below: A zinc oxide crystal attached to a supporting mesh of amorphous carbon.

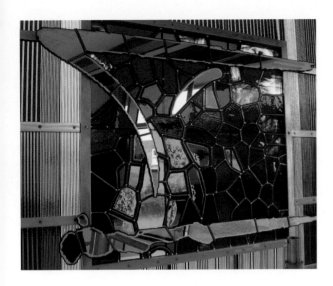

Stained glass designed by Tim Armstrong on display at the Nanoscience Centre. Using the Centre's logo as its focal point, Armstrong's work replaces the tip of an atomic force microscope with a sweep of calligraphy, symbolising the writing and recording of molecular structures.

the curious fact that the physical properties of structures with nanometre dimensions could be determined not simply by the material from which they were fabricated, but also from their size and shape. Remarkably, the Romans knew about this over 1,600 years ago. The Lycurgus Cup on display in the British Museum is made from glass impregnated with nanoparticles of gold and silver. The makers had worked out that if they ground down gold and silver sufficiently, the colour would change depending upon the light and the contents of the cup. The precise colour, from red to green, depended upon the size of the particles. Following the Romans, Cowburn and Welland fabricated patterns of nanoscale magnets, the shape and size of each magnet determining its specific magnetic properties. They speculated that such magnets could be equivalent to an electrical circuit in which the information was sent magnetically, and that perhaps there was a completely new way of transmitting at the nanoscale.

In another interdisciplinary research project, Welland and his co-workers pointed out that in a number of human diseases, such as Alzheimer's, the proteins essential for life assemble into long wire-like structures with diameters of only a few nanometres. They grew nanowires in the Nanoscience Centre using nanotechnology tools and measured their mechanical properties, discovering that these wire-like structures are amongst the strongest structures found in nature, similar in many ways to spiders' silk. Their remarkable conclusion was that, for a range of related diseases, the wire-like structures in human patients were very similar irrespective of the disease and, once formed, their strength made them very difficult to remove, explaining in part the challenges of reversing the damage done to the brain in patients suffering from Alzheimer's.

The Nanoscience Centre also provides space for interdisciplinary research as part of the Electrical Engineering Division of the Department and, as electrical engineering advances into new topical subjects such as healthcare, the facilities in the Centre are becoming increasingly important in linking the science and technology of the nanometre scale with the pervasive real-world applications of electrical engineering.

ELECTRICAL ENGINEERING

A striking building in West Cambridge, designed by the architects tp bennett LLP, houses the Electrical Engineering Division of the Engineering Department. It provides 4,800 m² of laboratory space, including state-of-the-art clean rooms, and accommodates 26 academic staff, around 70 postdoctoral research assistants and upwards of 150 postgraduate students. Funds for the building were scarce, but fortunately Harry Coles, a professor in the Department, noticed that a Corning Inc. Laboratory at Paliseau in France had installed a state-of-the-art clean room, but had decided for commercial reasons to close the facility in 2003. The enterprising engineers purchased the clean room structure and

Daping Chu, Director of the Centre for Advanced Photonics and Electronics.

equipment worth about £5 million at an online auction for a very small sum, had it dismantled and David Green got it moved to storage in Cambridge. It was then installed, at a fraction of the cost of a new facility, in the new building which was constructed over the next 18 months – much to the horror of the authorities who feared a disaster!

Several interdependent research groups are housed in the building, including the Centre for Advanced Photonics and Electronics (CAPE); the Centre for Photonic Systems; the Electronic Devices and Materials (EDM) Group; the Electronics, Power and Energy Conversion Group; the Graphene Centre; and the Hetero-Genesys Laboratory. It is impossible to include in a book of this length all the research activity in this extensive laboratory, and it has only been possible to select a few of the activities in three research groups.

CAPE was a consortium established in 2004 by Bill Milne and Bill Crossland to enable the University to address global engineering challenges, in partnership with companies of international standing and in an environment

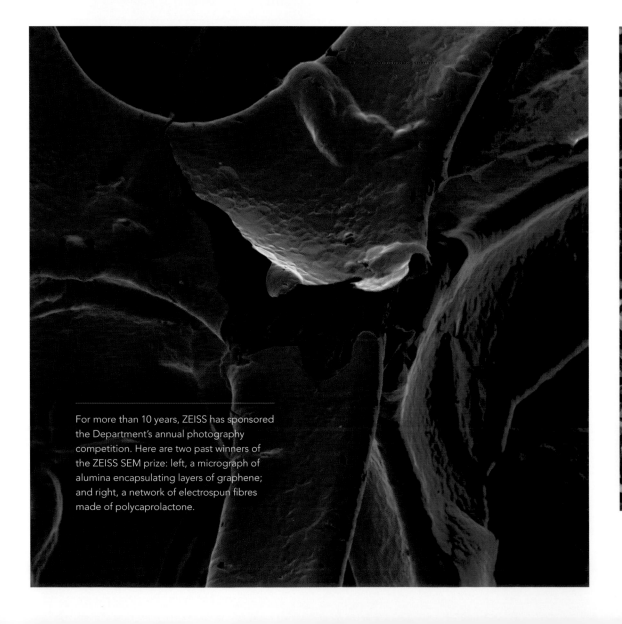

For more than 10 years, ZEISS has sponsored the Department's annual photography competition. Here are two past winners of the ZEISS SEM prize: left, a micrograph of alumina encapsulating layers of graphene; and right, a network of electrospun fibres made of polycaprolactone.

A Jaguar Land Rover heads-up display projects key driving information onto a windscreen. The technology was developed at the Centre for Advanced Photonics and Electronics.

of open innovation. It was expected that the research would be conducted in a new form of University–industry enterprise, with a small number of industrial partners and a global view of engineering enterprises, supporting and guiding the research programmes in the CAPE. The four original industrial members were Advanced Nanotech, Dow-Corning, Ericsson and ALPS (Japan), each contributing £500,000 per annum for the research programme. Apparently, the involvement of ALPS started with an informal conversation between Crossland and a director of ALPS in the University launch following the annual Boat Race on the River Thames. They were able to gather funds from the University's Science Research Investment Fund, the Department's funds and a loan from the annual funding available to the Electrical Engineering Division, enabling completion of the building.

Daping Chu is the present Chairman and Director of CAPE and Head of the Photonics and Sensors Group. A Fellow of Selwyn College, he is also Concurrent Professor of Nanjing University in China and his research interests are wide ranging, from semiconductor devices and materials to inkjet fabrication processes.

The Centre's primary responsibility, to translate science to technology for fast market implementation, is achieved by accessing world-leading expertise in every branch of engineering and science at the University, supported by collaborative work within the Electrical Engineering Division. One of the outcomes was expected to be important early-stage intellectual property; an example of research work moving rapidly into mass production can be found in the holographic head-up displays adopted by Jaguar Land Rover. These displays use laser holographic techniques to project information such as speed, location and navigation onto a vehicle's windscreen.

Currently, the CAPE partnership includes the Beijing Institute of Aerospace Control Devices (BIACD), CRRC Zhuzhou Institute, Disney Research and Jaguar Land Rover, along with ZEISS Microscopy as an Associate Partner. ZEISS donated two state-of-the-art scanning electron microscopes to CAPE, thus continuing a link with the Engineering Department that began in the 1950s with the

Andrea Ferrari, Director of the Cambridge Graphene Centre.

development of the Stereoscan by the Cambridge Instrument Company (later acquired by the company that is now ZEISS Microscopy). The donation enhanced CAPE's capabilities in its growing activity in nanostructured materials. Further activities of CAPE include managing the annual CAPE Acorn Postgraduate Research Award and the CAPE Acorn Fund for a Part IIB Project Award; organising a series of lectures, seminars and workshops (since 2004) and establishing strong links in China with academic institutions and industrial companies.

For many years, the focus of the Electronic Devices and Materials (EDM) group was on thin-film silicon materials, now widely used for fabricating transistors (that drive flat panel displays) and solar cells. This work diversified to include carbon-based electronic materials, such as diamond-like carbon for cold-

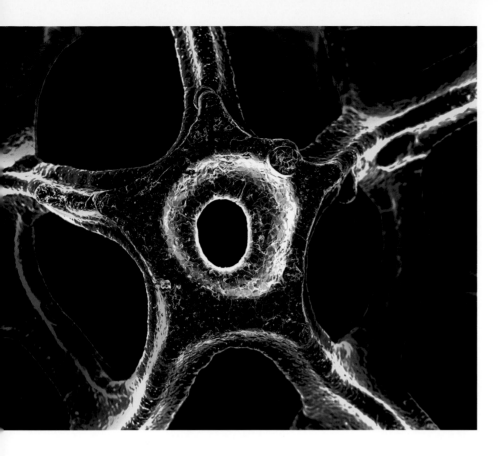

Electron micrograph of a free-standing graphene foam.

Opposite: Layers of graphene encapsulated by a three-dimensional shell of alumina.

cathode field emitters, carbon nanotubes and, more recently, graphene. Nanostructured carbons have proven to be well suited to actuating microelectromechanical systems such as microtweezers because of their very high Young's modulus of elasticity, while their high surface-to-volume ratio has opened the door to sensor applications such as micro-hotplate gas sensors.

There are also collaborative projects with commercial companies on diamond-like carbon for disc and media coatings and for the insides of plastic drink bottles, as well as applications in field emission devices. The Group designed and built its own equipment for manufacturing carbon nanotubes and founded Cambridge Nanoinstruments to manufacture growth systems commercially. Within 18 months, the company was bought by Aixtron, which is now the premier supplier of the equipment. The technique is also being utilised to grow monolayers of graphene up to 300 mm in diameter.

The Graphene Centre, in an annexe to the main building, is led by Andrea Ferrari. Its mission is to investigate the science and technology of graphene, carbon allotropes, layered crystals and hybrid nanomaterials. Industrial and academic partners promote innovative and adventurous research, with an emphasis on applications. The Centre was funded by the EPSRC Graphene Engineering programme. It is part of a Synergy Group with the University of Lancaster and the Graphene National Institute in Manchester, targeting its research on heterostructures and superstructures based on two-dimensional atomic crystals and their hybrids with metallic and semiconducting quantum dots. The facilities and equipment have been selected to promote alignment with industry by providing intermediate scale printing and processing systems where the industrial utilisation of inks based on graphene can be studied and two-dimensional crystals tested and optimised.

The unique properties of graphene necessary to achieve the goal of 'graphene-augmented' smart integrated devices on flexible/transparent substrates are being studied, particularly for their energy storage capacity – necessary for fabricating autonomous wireless-connected devices.

Innovations and Inventions:
Spin-Outs Changing the World

Across the whole of the Engineering Department, research projects led by senior academics and supported by research students and postdoctoral research workers are creating innovations and inventions that are addressing some of the major challenges of the twenty-first century and changing the world in which we live. Research groups are contributing to the built infrastructure, the environment, sustainability, communication and much else, and in many cases transferring their research to entrepreneurial enterprises. For this section, just some of the many research highlights in the Department have been selected and described in a style suitable for the non-specialist reader. It is by no means a comprehensive description of all the research in the Department but represents research that has stood the test of time as well as contemporary research in the Department that promises to change the world for the better.

CEDAR equipment was used on many well-known recordings and films, including *Star Wars* (1977).

DIGITAL MICROSURGERY: THE CEDAR PROJECT

A long-standing area of research in the Department that has been successfully combined with entrepreneurship is digital audio restoration. The topic was initiated in 1983 by a collaboration between the Founder and Head of the Signal Processing and Communications (SigProC) Laboratory, Peter Rayner (now Emeritus Professor of Engineering), and the British Library's National Sound Archive, who wished to carry out revolutionary digitisation and processing of their collections of old 78 rpm recordings and wax cylinders, removing pops, scratches and hiss from the sound – in other words digital 'microsurgery' on the signals stored in the grooves of historic recordings. Rayner's PhD student, Saeed Vaseghi, played a key role in taking the work forward. This led to innovative digital signal processing (DSP) techniques being developed and applied, with truly outstanding results that amazed listeners who were familiar with the limitations of analogue playback systems. DSP really could doctor the sound from these old recordings, to a seemingly magical extent!

It became apparent that the techniques developed in the laboratory could be commercially viable, and a spin-out venture called CEDAR Audio was proposed. Surprising as it may seem today, in the 1980s the Department and the University were

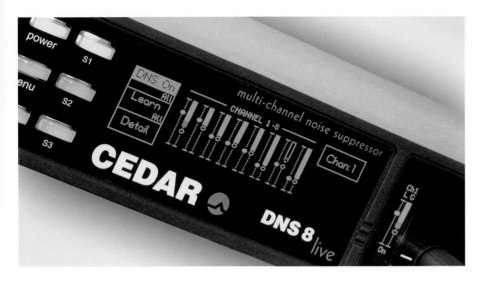

CEDAR technology removes the clicks and crackles from old audio recordings.

resistant to the idea of a company spun out from Departmental research, and a steep learning curve had to be traversed by the entrepreneurs. Eventually, a suitable arrangement was agreed, with the University taking a golden share in the company and the British Library funding the initial commercialisation.

At this stage, the current Head of the SigProC Laboratory and Rayner's former student, Simon Godsill, came on board as a founding staff member of CEDAR, to lead the DSP research and development and to make a commercially viable product. Real-time processing methods were developed on accelerator computer hardware and substantial algorithmic advances were necessary before a practical system could be realised in a form that would be attractive to customers. CEDAR's technology rapidly became the de facto industry standard, overtaking in capability and popularity the only competitor at the time, Sonic Solutions, a Californian company funded by George Lucas of *Star Wars* fame. (CEDAR later went on to remaster the original *Star Wars* soundtrack.) Godsill returned to complete his PhD, was appointed to the academic staff and later was promoted to a Professorship in Statistical Signal Processing. He retains a major interest in audio processing through his University research projects and interaction with CEDAR, and is a Fellow of Corpus Christi College.

CEDAR has since developed with increasing success, receiving numerous industry accolades, including an Academy of Motion Pictures Technical Achievement Award. The company has broadened its scope and now applies its methods to forensic audio for police forces, film sound and live audio feeds from television and radio broadcasts. CEDAR's innovative research-led approach to algorithm development and implementation, and its interactions with Godsill's group, make the technology transfer genuinely two-way, with the University group inspiring the adoption of state-of-the-art research ideas, particularly those applying Bayes' Theorem to advanced probability modelling of audio, and the company providing cutting-edge application ideas and datasets for future research. It is this model of University–industry interaction that has ensured that CEDAR has remained resilient through lulls in the economy and falls in the demand for audio, and that the Department's research has continued to be at the forefront of the subject.

BUILDING BETTER AND SMARTER UNDERGROUND INFRASTRUCTURE

The government and the civil engineering industry have combined to fund the Cambridge Centre for Smart Infrastructure and Construction (CSIC), with Robert Mair as principal investigator, complementing the Laing O'Rourke Centre for Construction Engineering and Technology at Cambridge. This Innovation and Knowledge Centre specialises in devising state-of-the-art instrumentation and monitoring, based on fibre optics, wireless sensor networks, computer vision and microelectromechanical systems. A highlight of the research is an innovative method of measuring strain in structures such as tunnels, shafts, retaining walls and piles, using distributed fibre-optic sensing to compute the strains and stresses. The data give new insights into the behaviour of a structure and help to reduce reliance on conservative safety margins. This provides greater assurance

Robert Mair, Sir Kirby Laing Professor of Civil Engineering. He was Head of the Civil Engineering Division until 2016.

of safety and reduces the quantity of construction materials needed and the time for completion. Major projects such as Crossrail, the largest European construction project, are applying Cambridge sensor technologies in 50 construction sites, and the very conservative civil engineering industry is taking notice of the benefits of advanced instrumentation and computer modelling.

Mair, the Sir Kirby Laing Professor of Civil Engineering at the Department, graduated in 1971 and worked in industry until 1998, except for a three-year period when he returned to Cambridge to study under the supervision of Andrew Schofield for a PhD on tunnelling in soft ground. In 1983, he set up a consultancy company, Geotechnical Consulting Group, with the avowed mission of introducing modern and innovative methods to the construction industry.

Mair's earlier work in civil engineering came during the Jubilee Line Extension project for London Underground, when twin tunnels under the River Thames had to be built in four places, along with several new underground stations. One of these was just across the road from the clock tower of the Houses of Parliament, which houses Big Ben, and featured an almost 40 m excavation to accommodate seven storeys below ground and seven storeys above ground. Instrumentation was cleverly introduced into the landmark tower and the buildings and ground surrounding it, with 7,000 monitoring points feeding into computers that analysed data to calculate where and when underground reinforcement was necessary. For safety, the tower could not move horizontally by more than 15 mm at a height of 55 metres, and this exacting specification was met by using the novel technique of compensation grouting pioneered by Mair and David Hight. Networks of horizontal tubes (*tubes à manchettes*) were permanently inserted between the ground surface and the tunnel wall, with holes along the length of the tube. A relatively weak cement and bentonite grout mix was injected into the tubes, creating ground heave and mitigating settlement during excavations, with further injections when indicated by the smart monitoring systems. This technique has been adopted widely around the world.

In the Crossrail project, compensation grouting is equally important, because the tunnels are located underneath the prime properties of Mayfair, Park Lane and Soho. In 2014, the CSIC group created a laboratory in a 100-year-old disused, cast-iron London Underground tunnel, installing monitoring devices to measure strain changes in real time during excavations of the gigantic Crossrail 12 m-diameter tunnel directly underneath the laboratory. Cast-iron tunnels are ubiquitous in the ageing London Underground network, much of it Victorian, and Cambridge technologies will enable engineers to gauge exactly what is happening in real time. For his services, Mair, a Fellow of the Royal Academy of Engineering and of the Royal Society, was appointed an independent crossbench peer in the House of Lords in 2015.

A team from the Cambridge Centre for Smart Infrastructure and Construction (CSIC) works late at night in the now-abandoned Post Office Tunnel near the new Crossrail project. CSIC researchers use innovative measurement systems to monitor the behaviour of the old tunnel as Crossrail construction takes place.

Abir Al-Tabbaa, Professor of Civil
and Environmental Engineering.

IMMORTALITY CRACKED:
SAVING OUR INFRASTRUCTURE

Infrastructure materials – concrete, mortars, cement and soils – form the bulk of bridges, tunnels, motorways, embankments and dams. Decades of neglect in the upkeep of ageing infrastructure has led to a shocking status quo cost of its repair and upgrade: about £40 billion a year in the UK and perhaps 10 times as much in the USA. The elimination of inspection, repair, maintenance and replacement of infrastructure assets will bring huge financial and environmental benefits.

In the Engineering Department, Abir Al-Tabbaa, Professor of Civil and Environmental Engineering and Director of the EPSRC Centre for Doctoral Training in Future Infrastructure and Built Environment, leads research on enhancing the resilience of present and future infrastructure by equipping its backbone with an 'immortality potion'. Al-Tabbaa is one of only three female professors in the Engineering Department. Having worked in industry and academia, her expertise spans across geotechnical, environmental and structural materials, with a focus on sustainability, durability and innovation.

Looking to natural systems for inspiration, the field of self-healing materials has emerged, in which systems are developed to look after themselves by sensing, responding to and repairing their damage without external intervention. The self-healing of construction materials in particular has proved to be far more tangible than initially thought, and it should be possible to transform a material like concrete from a mundane commodity into a 'living material' capable of looking after itself.

Over the past few years, Al-Tabbaa's group, in collaboration with the universities of Cardiff and Bath, has embarked on the challenge of developing structural and geotechnical infrastructure material systems that can repair myriad forms of complex damage, including external and internal cracking, corrosion of steel reinforcement and material expansion, to name but a few. Infrastructure materials are complex multicomponent and multilayered systems in which damage occurs at different length and temporal scales. The collaborative project takes its inspiration from the similarity with the intricate way in which complex skin systems repair damage.

Cambridge Simulation Gloves and Glasses give a young student an accurate sense of how an older person would perceive the world and use a product.

For small-scale damage repair, microcapsules filled with a healing agent have been developed for mixing within the bulk structural material. A crack ruptures some capsules, releasing their cargo, which fills and seals the crack. Mineral cargos, compatible with the material matrix, are being developed, as well as strains of bacteria that precipitate limestone. For larger-scale damage, shape memory polymers, in the form of plastic cables, are placed strategically inside the concrete, and these shrink when large cracks appear, thereby closing them. To deal with repeated damage, flow networks, analogous to the vascular blood system, have been developed to transport and dispense healing agents. Working with industrial partners, full-scale field trials were conducted on a major highways scheme in Wales and in the plant room of the new Dyson Building to validate the systems. Major advances in research have made possible commercialisation of some of the systems in the near future.

Above: Self-healing concrete demonstrating the cavity left behind after a microcapsule was broken and debonded from the material, with healing materials developing in the cavity.

DESIGNING FOR OUR FUTURE SELVES: INCLUSIVE DESIGN

It might be hard to imagine what a Ferrari Enzo and a jar of Duerr's jam have in common. Surprisingly, the Enzo was designed to be easier to get into than previous Ferraris, and Duerr's innovative 'easy open' lids were equally designed

Above: John Clarkson, Director of the Engineering Design Centre, with two Roberts radio 'revivals' of old-style designs. The designs encourage older users familiar with earlier analogue controls to use the new digital technology.

Simple calculations, Discrete Event Simulation (DES) and a LEGO Mindstorms model – 'LEGOline' – are all used to illustrate the challenges faced by engineers designing real-life systems. Alexander Komashie shown with LEGOline.

Patient safety report that was distributed in an eye-catching blood bag across the NHS in 2003. The system-wide design-led approach was landmark in improving the safety of healthcare.

with better access in mind. Both changes were in direct response to the physical challenges faced by many older people, and both resulted in better products for all potential customers.

The world's demographics are changing, with longer life expectancies and a reduced birth rate resulting in a higher proportion of older people. Yet with increasing age comes a general decline in capability, challenging the way people are able to interact with the 'designed' world around them. The Cambridge Engineering Design Centre has worked with the Royal College of Art to address this 'design challenge', leading to a new definition of inclusive design as 'the design of mainstream products and/or services that are accessible to, and usable by, as many people as reasonably possible', and finding ways in which business and designers can be better equipped to commission and design inclusive products.

The Cambridge team sought to answer a number of key questions, namely: How do people's sensory, cognitive and motion capability vary with age? How might this be measured and/or simulated with reference to product interaction? How might such knowledge be transferred to business and designers to facilitate better design? In answering these questions, the team developed a design toolkit, and realised what was by now obvious, that inclusive design was simply better design.

When the demands made by a product in a given situation exceed the capability of the individual to respond, that person is likely to be excluded from using the product. There is likely also to be a significant number of people who, for the same reasons, find that product difficult or frustrating to use. Any subsequent improvements to the design to exclude fewer people are then very likely to make it easier and less frustrating to use for the majority. The resulting online *Inclusive Design Toolkit* has enabled designers from all over the world to access information and tools for better design, and training workshops have been provided for many leading international companies. In addition, education resources have been developed with teachers to deliver the same insights and tools in the classroom.

The Engineering Design Centre was founded in 1991 by Ken Wallace, Mike Ashby and David Newland. For more than 25 years, it has undertaken fundamental and applied research to generate knowledge that improves the design process. John Clarkson, Professor of Engineering Design and Director of the Engineering Design Centre, read engineering and stayed in the Department to complete a PhD under the supervision of Paul Acarnley before joining PA Consulting Group as a systems engineer. He was elected a Fellow of the Royal Academy of Engineering in 2012 and is a Fellow of Trinity Hall.

MATERIAL FACTS

Ashby and David Cebon collaborated in computer-based materials teaching in the Engineering Department for several years, developing the Cambridge Materials Selector software for plotting materials property charts and using them for selection, and more generally for teaching undergraduates materials engineering – an initiative strongly supported by successive Heads of Department, Broers and Newland. They recognised that the resources developed for education had the potential for commercial exploitation, and their small team was encouraged by Broers to work in an attic room in the Department in return for a small rent while a viable product was under development. Cebon writes: 'This made all the difference in our ability to launch a company.' Initially, the software was licensed to customers in education and industry with the help of Richard Jennings from the Wolfson Industry Liaison Unit (now Cambridge Enterprise), and it was soon apparent to the inventors and the University that there was an opportunity for worldwide exploitation.

An example of an Ashby chart, which is used to select materials.

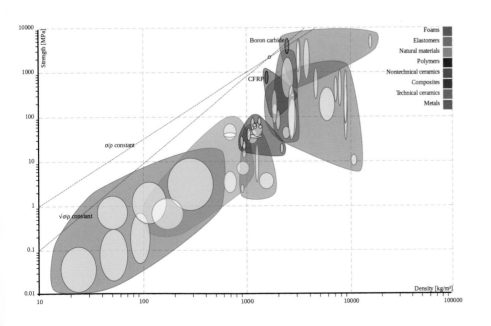

In 1994, Cebon and Ashby founded Granta Design as a spin-out from the Department, and the company expanded rapidly on the strength of its intellectual property; an extensive compilation of materials data; its materials property chart, commonly known as the Ashby chart; methods for presenting and analysing materials data; and the performance index concept – using property charts to identify optimal materials choices for a specified engineering application.

A key event in the company's history was the investment in 2000 from ASM International, which became a strategic partner and shareholder together with the University of Cambridge, the founders and the staff of Granta Design. The company continues to maintain a strong focus on innovation, ensuring that it is able to provide new technologies and products that meet changing customer needs. The Education Division of Granta develops teaching resources and coordinates International Materials Education Symposia in the USA, Europe and Singapore. Its core product, the CES EduPack, comprises teaching resources for university-level materials teaching over a wide range of related engineering, science and design courses. The Commercial Division develops, sells and supports 'Granta MI', the industry standard for enterprise materials data management, used by companies in materials development for aerospace, defence, energy, automotive, consumer electronics and medical devices, as well as in commercial research organisations.

Today, Granta is a world leader in materials information technology, employing 140 staff in an international organisation focused solely on its niche field. Its materials information management systems interface with leading CAD, CAE (computer-aided engineering) and PLM (project lifecycle management) systems, and have become the tools of choice for major enterprises across the world. Granta's educational products are now licensed by over 900 universities and colleges, and are used by many thousands of students every year. The company has its headquarters at Rustat House in Cambridge, where 80 personnel are employed.

The founders maintain that their success was enabled by the positive attitude and encouragement of both the Department and the University, and that its products are changing the way that materials information is used to make better, safer, greener products worldwide. Mike Ashby, Emeritus Professor of Materials, works in the Engineering Design Centre in the Department. He was Professor of Applied Physics at Harvard University until 1973, when he was appointed to a chair in the Engineering Department. He is a Fellow of the Royal Society and of the Royal Academy of Engineering and was appointed CBE in 1997. David Cebon, Professor of Mechanical Engineering and Fellow of the Royal Academy of Engineering, leads the Transport Research Group in the Department's research theme of energy, transport and urban infrastructure. He is Managing Director of Granta Design Ltd.

Opposite: A robotic unicycle used in student projects and research to develop new learning algorithms to achieve robust balance and control.

Below: Cambridge Enterprise was formed to help students and staff commercialise their expertise and ideas.

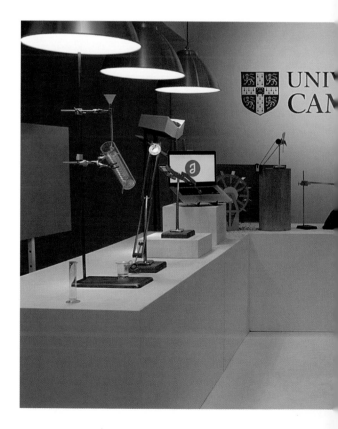

MECHANICAL SCIENCES TRIPOS PART II, 1897

Investigate the equations of motion of the governing masses in some form of Watt's centrifugal governor and find an expression for the period of the oscillations about the state of steady motion. Point out in what circumstances the oscillations will degenerate into hunting and shew how this fault may be prevented.

ALMOST EVERYTHING UNDER CONTROL: FLYING IN THE FACE OF UNCERTAINTY

The great nineteenth-century savants James Clerk Maxwell and Edward Routh laid down the analytical framework for the design and performance of feedback control in a classic paper on Watt's governor and an Adams Prize essay on stability criteria. Cambridge's presence in the early developments in control theory was assured (readers are invited to solve the Tripos question in the box above). Control theory or control engineering is now recognised as a discipline that cuts across almost all branches of engineering, even extending beyond the conventional boundaries into biology and beyond.

The first lecture courses on control engineering in Cambridge as a discipline in its own right were delivered by G D S MacLellan in 1946, just before he spent a sabbatical year at MIT's Servomechanisms Laboratory. From there he wrote to Baker: 'MIT was certainly the right place for me to choose to come to, but the scale of its activities takes some getting used to. It is in almost the same relation to Cambridge as Cambridge is to Oxford as regards Engineering!' Another lecturer, Robert Macmillan, started research in control engineering and published *An Introduction to the Theory of Control in Mechanical Engineering* in 1951 before resigning his lectureship in 1956 to take the Chair of Control Engineering at Swansea University. Control engineering received a further boost in 1953 when Baker recruited John Coales, who was Director of Research at Elliott Brothers, to establish a Control Group within the Department and a postgraduate course. Coales established a thriving group and was also significant outside the Department, becoming the President of the International Federation of Automatic Control (1963–1966) and then President of the IEE (1971–1972).

Tom Fuller added to the prominence of the group with his work on mathematical control theory, such as his results on optimal control laws for simple systems with a saturated input showing that these might require either an infinite number of switches in a finite time (now called the Fuller Phenomenon), or that there did not exist a linear control law that stabilised a triple integrator. Alistair MacFarlane, elected Professor of Engineering in 1974, established a Division of Control and Management Systems in the Engineering

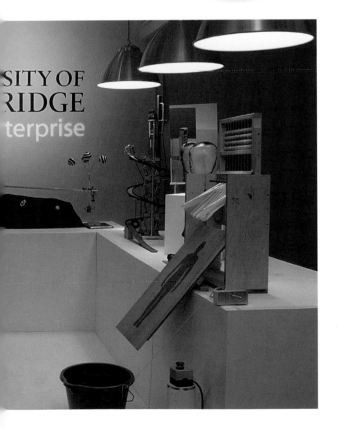

Department. His research was concerned with feedback control of systems with many inputs and outputs, using frequency domain methods and complex function theory, which he then embedded in computer-aided control system design software. In 1984, he formed the Information Engineering Division, which included the Signal Processing and Speech Research Groups. He resigned from his professorship in 1988 to become the Vice-Chancellor of Heriot-Watt University and was knighted in 2002.

Keith Glover was elected to the Professorship in 1989. He enhanced the reputation of Cambridge as a prominent centre for control research with his celebrated work on model approximation and H-infinity control, recognised in 2001 by the highest honour available to a control engineer: the IEEE Control Systems Award. This work addressed the robust control problem where the performance of a feedback system is to be designed to be robust in spite of the presence of uncertainty in the system being controlled. Although based on the abstract mathematics of functional analysis, the results were made theoretically accessible and computationally tractable. They are now incorporated into control system analysis and design software packages used by industry and universities around the world. One paper received the IEEE Baker Prize for `the most outstanding paper reporting original work in the Transactions, Journals and Magazines of the Societies or in the Proceedings of the IEEE' (this includes approximately 100 journals). This paper was authored by Doyle, Glover, Khargonekar and Francis, with the co-authors from Caltech and the universities of Minnesota and Toronto. Glover's more recent work has included exploiting this design methodology in flight control which has included flight tests on a Harrier aircraft set up for advanced control assessment and gave excellent results. Most recently, he has been collaborating with

A Team Penske IndyCar Series racing car; Malcolm Smith's Inerter technology is licensed to Penske.

Nick Collings on internal combustion engine management to reduce toxic emissions and fuel consumption and to improve performance.

In due course, Glover was joined by two new lecturers, Malcolm Smith and Glenn Vinnicombe, who both returned from faculty positions in the USA, having previously gained their PhDs in the Control Group. The principal focus of their work was also robust control, knowing that the closed-loop performance of a feedback system can be extremely sensitive to small uncertainties. A measure of the uncertainty (the gap metric) is an effective design and analysis tool. The Vinnicombe metric (the nu-gap) is considered the most elegant version of this metric.

In 1991, Smith started work on active suspension systems for Formula 1 racing cars, but when these were made illegal in 1994 he moved on to study passive suspensions from a fundamental standpoint, noting that passive circuit synthesis for electrical RLC (resistor/inductor/capacitor) circuits could not be applied to spring/mass/damper mechanical suspension systems because the mass-to-capacitor analogy is not exact (electrical current is proportional to rate of change of voltage across a capacitor, whereas mechanical force is proportional to rate of change of velocity of a mass relative to ground). In Smith's conceptual invention – the inerter, a device with two terminals – the force is required to be proportional to the difference in the acceleration of the two terminals. A practical device developed by him working with engineers at McLaren Racing was first used in Raikkonen's winning car in the Spanish Grand Prix in 2005; McLaren subsequently won 10 of the remaining 15 races in the season. McLaren kept the component confidential and referred to it as a J-damper, which led to speculation in the press on the nature of the device and a spying allegation against another team that was unable to determine the function of the inerter in spite of unauthorised access to the technical drawings! Inerter technology is now licensed to Penske Racing Shocks and used extensively in F1 cars, IndyCars and elsewhere.

The dramatic improvements in the speed of computers and algorithmic advances enabled control methods that had been restricted to slow chemical processes to be applied to a wide range of systems. Jan Maciejowski and his students have been central to the algorithmic developments, with studies including aircraft control in the face of failures and the docking of space craft. The quantitative methods developed for control theory have also spread to systems biology. Vinnicombe, Lestas and Paulson published a remarkable result in *Nature* (2010) establishing limits on the accuracy of the regulation of the concentration of a 'target' molecule in a cell. Mechanisms for this often

Jan Maciejowski, emeritus Professor of Control Engineering and previously Head of the Information Engineering Division.

Rodolphe Sepulchre, Professor of Engineering in the Control Group, researches nonlinear signals and systems.

involve a 'mediator' molecule whose birth rate depends on the number of target molecules and no other information. By developing mathematical tools that merge control and information theory with physical chemistry, they showed that the standard deviation of the concentration of the target only decreases with the fourth root of the birth rate of the mediator, leading to the conclusion that 'making a decent job is 16 times harder than making a half decent job', no matter how complex the regulating mechanism. This goes some way to showing why noise is so prevalent in the cell.

Following the retirement of Keith Glover, Rodolphe Sepulchre was elected to the Professorship of Engineering (2012), and he diversified the research by applying control methods to the understanding of neuroscience, building links with Wolpert's group in Computational and Biological Learning and other research groups in the wider University. Recently, he has been joined by two new lecturers: Fulvio Forni (control of nonlinear and hybrid systems) and Timothy O'Leary (theoretical and computational neuroscience and physiology of the nervous system).

COMBUSTION TO CAMBUSTION

In the 1930s, Harry Ricardo worked not only on the development of the diesel engine but also on controlling its emissions. The role of these and other emissions in damaging air quality in cities was recognised in the early 1970s and, since alternative prime movers were a distant prospect, the development of 'cleaner' vehicles became a major driver

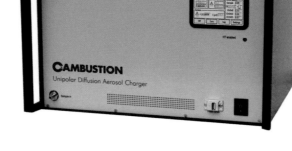

in automotive technology. In the 1980s, Nick Collings, Professor of Applied Thermodynamics, and his research team invented a sensor for unburnt fuel from an internal combustion engine, with a response time of one millisecond – 10,000 times faster than any of the conventional sensors fitted in cars, and fast enough to identify individual cylinder firing events.

Although this work was fundamental research, Collings recognised the possibility of commercial exploitation in the form of an instrument based on the sensor, and in 1987 a private company, Cambustion Ltd, was spun out by Collings, together with three of his research students and two undergraduates whose final-year projects were linked to his research, to produce a fast-response Flame Ionisation Detector (FFID) for hydrocarbons. Collings explains: 'The vast majority of the particles we inhale in an urban environment are from diesel or gasoline engine fumes. Unless a vehicle can pass the increasingly rigorous limits placed on the emissions of unburnt fuel, oxides of nitrogen, carbon monoxide and particulate matter, it just cannot be marketed.'

Right and below: Cambustion products have been sold in 26 countries around the world, where they are used for both fundamental combustion research in universities and to improve the emissions performance of engines by vehicle manufacturers. Below: a unipolar diffusion aerosol charger. Right: equipment for measuring the size of aerosols.

Opposite: Victoria Cronin, then a PhD student in the Centre for Sustainable Development, stands with local children while researching the sustainability of alternative approaches to slum upgrading in Kibera, Kenya.

In the mid-1980s, as limits on permitted emission levels became even more stringent, Collings decided to explore a fundamental engineering problem in internal combustion engines – because the emissions were emitted mainly during short transients occurring during cold starts and vehicle accelerations, they lasted just fractions of a second whereas the equipment used to measure them had a response time of 10 seconds. Following the success of the FFID, ultrafast sensors, with the ability to detect not just unburnt fuel but also the oxides of nitrogen and other noxious gases as well as particulate matter, were invented and successfully marketed by Cambustion.

Cambustion now employs 65 staff in Cambridge and supplies its unique instruments across the world through a network of agents, while a specialist team offers consultancy services to equipment manufacturers using instruments to develop 'cleaner' vehicles. Although fumes from individual vehicles are decreasing as engines become cleaner, the volume of traffic continues to grow, and ambient air quality standards are becoming very much more demanding. Recently, in collaboration with Keith Glover, a new theme of engine control has emerged within the Engine Research Group and a series of projects is underway, where the synergism of advanced control techniques and sophisticated measurements is being explored.

THE ENGINEER IN CONTEXT: THE CENTRE FOR SUSTAINABLE DEVELOPMENT

What can modern engineers do to embrace and implement sustainable solutions to the many and varied technological problems they face? A busy practising engineer can find some of the answers in a series of books edited by the Engineering Department's Centre for Sustainable Development, including *Sustainable Infrastructure: Principles into Practice*, *Sustainable Water*, *Sustainable Buildings* and *Sustainable Transport*. This thriving cross-disciplinary research group confronts a diverse and challenging range of research themes, where vested interests are not necessarily served but sustainability is enhanced and the interests of whole communities are addressed. The social value of infrastructure, the impact of pre-construction delays in large dams, resilience and decision-making in the aftermath of disaster, carbon implications of contrasting forms of travel, and the combined impact of social and structural vulnerability in an earthquake-prone region are some of the topics studied in the Centre.

The Centre for Sustainable Development was established in 2000, with funding from the Royal Academy of Engineering for the first post in the UK of Professor of Engineering for Sustainable Development, to which Peter Guthrie was appointed. This built on a Visiting Professor position held by Guthrie in 1999, which was then filled by Charles Ainger. The Cambridge MIT Institute (CMI) provided funding for the first five years of an MPhil course in Engineering for Sustainable

Development. The course commenced in 2002 with Dick Fenner as Course Director and an intake of 14 students, growing rapidly in subsequent years to more than 50 students. During its 15 years of operation, the course has attracted professionals from across the world, dedicated to making sustainable development an integral part of their career. Building on technical foundations, the course encourages engineers to think differently about problems in a world where the constraints (and opportunities) for engineering practice are ever more complex.

Over 500 alumni now work as academics and as professional engineers in consultancy and contracting, applying the principles of sustainability in their everyday work, many with key roles in international development agencies such as the World Bank and Asian and African Development Banks. Research in the Centre for Sustainable Development provides a beacon for new approaches to messy problems in the built environment, which could provide sustainable yet workable solutions for the future.

The Centre has contributed to a number of projects aimed at reducing energy usage, including a review of energy efficiency in the NHS leading to a revision of the Health Technical Manual HTM 07; the dissemination of

A village engineer repairs a pump in rural Malawi, taken during student Anthony Rubinstein-Baylis' charity work teaching in the village.

retrofit measures in cities for improved energy performance; energy efficiency improvement in commercial building stock; and an investigation into the effectiveness of solid wall insulation of domestic buildings.

On the theme of water conservation, the Centre has a global outlook, focusing particularly on water, sanitation and sustainability issues in developed and developing countries. Areas of research include the impact of climate change and water treatment plant operation; life-cycle analysis of ecological sanitation systems; the implementation of household water systems in India and Nepal; an evaluation of business models for resource recovery and reuse of solid waste and wastewater in developing countries; and thermal heat recovery from wastewater. The Centre has also investigated the sustainability of water distribution systems in partnership with Yorkshire Water, evaluated the wider benefits of adopting green infrastructure solutions in sustainable drainage systems for urban flood risk management and helped to develop an evidence base explaining why rural water supplies in Africa frequently fail.

Frank Fallside was a pioneer in the area of information technology and speech processing. He established the MPhil course in Computer Speech and Language Processing in 1985.

TALKING TO A COMPUTER: THE MACHINE INTELLIGENCE LABORATORY

Today, the Machine Intelligence Laboratory has four academics – Steve Young, Phil Woodland, Mark Gales and Bill Byrne – working on speech and language processing and the research topic has advanced with each passing decade since the 1970s. It now includes large vocabulary recognition, spoken dialogue systems, speech synthesis, statistical machine translation and automated language testing, and it is changing the world of human–computer interaction. Collaborations with industry include the industrial giants IBM, Google and General Motors, with two spin-out companies: Phonetic Arts developing speech synthesis for the games industry and VocalIQ providing spoken dialogue for personal assistant applications.

In the 1970s, Frank Fallside was interested in linear prediction and inverse problems, and knew that human speech production could be modelled as a variable-sized tube excited by a mixture of pulses and white noise. He believed that it should be possible to estimate the shape of the tube used to generate a sample of speech – if this shape could be displayed on a screen, deaf children could learn to speak more clearly. Fallside's group had only a

From left: Mark Gales, Phil Woodland and Steve Young of the Machine Intelligence Laboratory.

PDP8 computer with eight kbytes of memory, so it had to build a real-time deaf training aid for system trials. The research grew and by 1984 speech processing was supported by EPSRC and two of Fallside's able students, Young and Woodland (now Professors of Information Engineering), who joined him as lecturers in speech processing. With Fallside at the helm, they became the acknowledged worldwide leaders in speech technology.

Fallside's intuitive feel for new areas of research led him to neural networks, thus creating novel applications of computer-based systems, in handwriting recognition, image processing and computer vision. The expanded group was renamed the Speech, Vision and Robotics (SVR) Group. Its new laboratory was named after Fallside following his tragic death in 1993, and the group is now called the Machine Intelligence Laboratory.

In speech processing, the Laboratory is distinguished internationally for its contributions to the development, by Young and Woodland, of automatic speech recognition. The core algorithms are embodied in a software system, that became the de facto standard in speech processing and formed the basis for the Cambridge Large Vocabulary Speech Recognition System that set the pace in the subject throughout the 1990s. In fierce competition with such institutions as IBM and the Stanford Research Institute International, the Cambridge system came in first place on 10 occasions in 13 evaluations carried out over a decade by the US National Institute of Standards and Technology. Twenty-five years after it was first created, the software is still widely used; it was the catalyst for Entropic, a University spin-out sold to Microsoft in 1999, and it still influences systems powering today's speech-based personal assistants in Google, Microsoft and Apple products. Fallside, not a demonstrative man, would have smiled with quiet pleasure to see what his progeny had achieved.

Zoe is a virtual talking head that can project human-like expression of emotions, developed by Roberto Cipolla's team in the Department of Engineering and researchers in Toshiba's Cambridge Research Lab.

BREAKING A WORLD RECORD: THE BULK SUPERCONDUCTIVITY GROUP

In 2014, the Bulk Superconductivity Group in the Engineering Department announced in *Superconductor Science and Technology* a remarkable experimental result: the highest trapped magnetic field ever generated by a bulk superconductor, 17.6 Tesla, in a two-sample arrangement of gadolinium barium copper oxide (GdBCO). This was greater than the previous best recorded trapped field of 17.24 Tesla achieved in Japan in 2003. The single-grain bulk superconductors, each of 24 mm diameter, were fabricated by a modified melt processing technique developed at Cambridge, and were reinforced by 'shrink-wrapping' a stainless steel ring around each single grain to increase its strength. The generated field contained an energy density that was around 11% of that of TNT, which highlights the remarkable potential of these technologically important materials, and this result reinforced worldwide interest in developing devices of bulk high-temperature superconductors in general.

Conventional superconductors are materials that conduct a direct current with zero resistance when cooled below a critical temperature close to absolute zero (-273 °C) using liquid helium. The phenomenon was discovered by Heike Kamerlingh-Onnes in 1911, and remained a subject of scientific investigation until Alex Müller and Georg Bednorz demonstrated in 1986 that superconductivity could be observed in certain materials, described by the misnomer 'high-temperature superconductors', when cooled with relatively low-cost liquid nitrogen (-196 °C). Engineers across the world viewed this result with excitement because of the possible practical applications of these extraordinary materials.

Circulating currents, induced in a bulk superconductor by the Lenz–Faraday effect, create a magnetic field that is related to the amount of current carried in the sample, which is typically hundreds of times greater than the current carried by copper at low temperature. This differs fundamentally from the mechanism by which permanent magnets, such as NdBFe (neodymium boron iron) or SmCo (samarium cobalt), generate magnetic field, which is by the alignment of unpaired spins, and is therefore limited to a maximum of around 1.7 Tesla. The field generated by bulk superconductors, on the other hand, is determined only by the size of the induced current and the length scale over which it flows, so bigger samples generate increasingly larger fields without a practical upper bound.

Below: A bulk superconductor levitates above a magnet.

From left: Mark Ainslie, David Cardwell and John Durrell of the Bulk Superconductivity Group.

The Engineering Department's Bulk Superconductivity Group is led by David Cardwell and includes Yunhua Shi, who was responsible for developing the melt process and fabrication technique, based on 20 years of research in the group. The Cambridge team, led by John Durrell, collaborated closely with the Boeing Company and the US National High Magnetic Field Laboratory in Tallahassee, where measurements were made using a novel high magnetic field facility.

Low-temperature superconductors cooled by liquid helium are used routinely in MRI devices, which are established as a major diagnostic tool in hospitals around the world, so the phenomenon of superconductivity is already in use in the public domain. Applications that are particularly relevant to high-temperature bulk superconductors include transport systems based on levitation, energy storage in a mechanical flywheel, medical equipment such as magnetic resonance imaging, magnetic separation and smaller, lighter electrical motors and generators.

The Bulk Superconductivity Group. Front row, from left: John Durrell, Archie Campbell, Mark Ainslie, David Cardwell and Yunhua Shi; Back row: Tony Dennis and Kysen Palmer.

POWERING THE FUTURE

Research on power generation and electrical energy transfer technologies, based on the physics of semiconductors allied to information systems, is conducted in the Engineering Department under the leadership of Gehan Amaratunga (1966 Professor of Engineering, Head of the Electronics Power and Energy Conversion Group and Fellow of the Royal Academy of Engineering) and Florin Udrea (Professor of Semiconductor Engineering, Fellow of the Royal Academy of Engineering and a Member of the Board of Cambridge Enterprise).

Applications of this University research have been found in intelligent chips for power supplies, LED driving circuitry, hybrid and electric vehicle inverters and motor control inverters. Amaratunga is also responsible for research on inorganic thin-film solar cells on plastic substrates, solar-powered LED lighting and hydropower integrated circuits, while Udrea carries out pioneering research in CMOS-based gas sensors using micro hotplates, infrared emitters and detectors, temperature sensors and flow sensors. The transfer of this technology to commercial enterprises is changing the world of power consumption.

The commercial potential of their research was quickly recognised by Amaratunga and Udrea, who founded Cambridge Semiconductor (CamSemi) in 2000; within a few years, the company became a leader in power management systems integrated into single chip-based controllers. It was named as the Spin-out of the Year in 2009 and recognised shortly afterwards as the fastest-growing semiconductor company in the UK. When sales reached one billion units, CamSemi was sold to Power Integrations, a Silicon Valley company.

Emboldened by this success, more entrepreneurial companies were founded, including Cambridge CMOS Sensors, which specialises in environmental monitoring using miniaturised metal-oxide gas sensors and infrared technology. The company was founded by Udrea, Bill Milne, and Julian Gardner from Warwick University, and it gained the National Microelectronics Institute Product of the Year Award, with Udrea, now Chief Technical Officer, serving as Chief Executive Officer. The founders then joined Nat Edington and Robert Swann in a venture using technology developed at the University of Cambridge for smart homes and mobile phones. Cambridge CMOS Sensors was acquired in June 2016 by Austria Micro Systems (AMS) and it is considered as one of the most successful University spin-outs in engineering and physical sciences to date.

Below right: Gehan Amaratunga (right) receiving the Royal Academy of Engineering Silver Medal from John Browne in 2007. The Silver Medal is awarded for an 'outstanding personal contribution to UK engineering by an early to mid-career engineer resulting in market exploitation'.

Below: Florin Udrea received the same prestigious Silver Medal five years later.

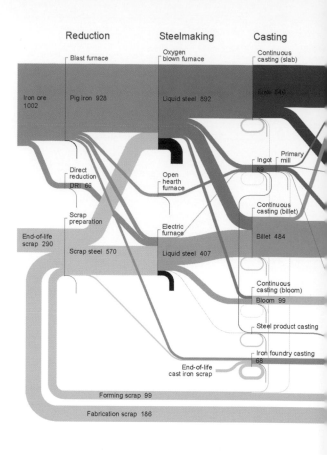

The research group is driven by a philosophy of transferring research rapidly towards commercial exploitation. Amaratunga has founded Nanoinstruments, acquired by AIXTRON AG, a leading provider of deposition equipment; Enecsys, which makes solar power modules; and Wind Technologies, specialising in generators for wind turbines. Recently, Udrea, Amaratunga, Tanya Trajkovic, Nishad Udugampola and Vasantha Pathirana have founded yet another company, Cambridge Microelectronics, which is working with silicon power semiconductor devices based on a unique lateral high voltage technology suitable for AC–DC converters and LED drives. This technology is capable of competing with gallium-nitride-based processes.

Udrea, in collaboration with Amaratunga and Milne, has also established within the Engineering Department a High Voltage Microelectronics and Sensors Laboratory, which specialises in power devices and sensors using smart power and micromechanical systems technology. The group has a prominent presence on high voltage device research at the International Electron Device Meetings and the International Symposium on Power Semiconductor Devices and Integrated Circuits.

USE LESS ENGINEERING

Most environmental problems turn into requirements for more energy. In meeting the need for ever more scarce resources more energy is used, for example in desalinating water, digging deeper for minerals or irrigating and fertilising to increase agricultural yields. A low-carbon supply of energy that scales to the current use of fossil fuels remains elusive. The challenge for humankind of responding effectively to the dangers of excessive climate change is to use less energy. The challenge for industry, where most energy is used to make bulk materials, particularly steel and cement, is to use less material.

The Use Less Group – with Julian Allwood, Professor of Engineering and the Environment, as Head of the Group – has three strategies for contributing to living well while using less energy and material. A portfolio of 'whole systems' analyses of the stocks and flows of materials and energy in use at national or international scales has been developed; presented using Sankey diagrams and other graphics, it identifies opportunities for change, highlights how different mechanisms interact, and exposes inefficiencies.

The Group, with University partners, is funded by BP to develop dynamic graphical presentations of future resource usage. The resulting FORESEER software, developed by a spin-out company, Foreseer Ltd, has been used to explore conflicts between energy and water policy in China; to evaluate UK

Below: Julian Allwood, Professor of Engineering and the Environment.

Below right: Researchers in Allwood's laboratory.

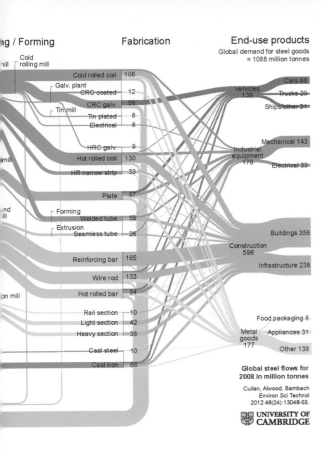

Global demand for steel goods = 1088 million tonnes

Global steel flows for 2008 in million tonnes

Cullen, Alwood, Bambach
Environ Sci Technol
2012 46(24):13048-55.

UNIVERSITY OF CAMBRIDGE

Above: An example of a Sankey diagram visualising the transfers between processes involved in making steel goods.

Above right: The flexible ring-rolling machine in Allwood's laboratory, which saves energy and materials by subtly easing cold metal feedstock into its final shape through many small incremental actions rather than using high force, cutting or heat.

climate change plans; to evaluate new investment decisions; and to examine the opportunity for recycling scrap for manufacturing steel.

Another Use Less project is to find ways to live well while using less material. Investigations into structural design discovered that multistorey buildings use double the steel required to meet safety standards, because it is cheaper to use more steel than pay the cost of designing and constructing efficiently. In car manufacturing, 50–60 per cent of the sheet metal is cut off during blanking or after stamping. Even more remarkably, in the aerospace industry, an enormous amount of swarf is created when parts are machined, which leaves the aeroplanes as merely a by-product that accounts for just 10 per cent of material! The Use Less Group works on creating change: different business models, new approaches to design and the development of an economic and policy context for decisions.

The third strand of Use Less engineering is designing processes that waste less metal. A prototype flexible spinning machine that makes axisymmetric products very near to net shape has been built in the Department, and another prototype is being commissioned at a Catapult Centre (government-sponsored centre for innovation). A new flexible ring-rolling machine is being designed to make annular components, required in turbines and in the oil and gas industry, close to final shape, reducing machining swarf to a minimum. A laser 'un-printing' machine that removes toner from paper, allowing re-use, has led to a second spin-out company, Reduse Ltd, where a full-scale prototype has been built.

Finding technologies that eliminate environmental problems is politically attractive and generously funded with over-optimistic predictions. The reality of dealing with climate change is to use less. Use Less engineering, together with a network of other disciplines, government and private sector partners, has become a major topic of research.

ENGINEERING THROUGH LEARNING: REVERSE-ENGINEERING THE BRAIN

The ease with which humans move arms, eyes, even lips when speaking masks the true complexity of the control processes involved. This is evident when attempts are made to build machines to perform human control tasks. While computers can now beat grandmasters at chess, no computer can yet control a robot to manipulate a chess piece with the dexterity of a six-year-old child. Moreover, movement is the only means of interacting with the world, whether foraging for food or attracting a waiter's attention! Indeed, all communication, including speech, sign language, gestures and writing, is mediated via the motor system.

Taking this viewpoint, the purpose of the human brain is to use sensory signals to determine future actions. Thus the goal of the research of Daniel Wolpert is to understand the computational principles underlying the control of human movement. Trying to understand how the brain controls movement requires techniques from areas such as machine learning, control theory, mechanics and signal processing – areas in which the Engineering Department has tremendous strengths. In addition, the experimental study of human motor control has advanced rapidly, with the development of robotic interfaces and virtual reality systems that allow for precise experimental control over sensory inputs and task variables. Using such methodology, the Group has shown that the brain builds up statistical models of the world and human bodies which it then uses within a Bayesian framework for control.

The Computational and Biological Learning Laboratory was formed by Wolpert and Zoubin Ghahramani, with the aim of applying engineering principles to reverse-engineer how the brain learns and to develop new machine learning techniques that could learn through experience. Wolpert and Ghahramani share an interest in uncertainty and Bayesian decision theory, but from diametrically opposite views; Wolpert seeks to understand how the brain learns in the face of uncertainty, while Ghahramani leads the Machine Learning Group and seeks to develop algorithms that could operate as well as and ultimately better than the human brain.

How can machines and humans learn from experience? Probabilistic modelling provides a framework for understanding what learning is, and has therefore emerged as one of the principal theoretical and practical approaches for designing machines that learn from data and understanding how brains learn through experience. Thus the probabilistic framework, which describes how to represent and manipulate uncertainty concerning models and predictions, has a central role in scientific data analysis, machine learning, robotics, cognitive science and artificial intelligence.

Zoubin Ghahramani – Professor of Information Engineering, Head of the Machine Learning Group and Fellow of St John's College, elected a Fellow of the Royal Society in 2015.

Above right and right: A Cambridge-led team revealed the engineering behind the Dambusters raid to blow up German dams during the Second World War in the Channel 4 documentary: *Dambusters: Building The Bouncing Bomb.*

Daniel Wolpert (right) and his team. Wolpert is 1875 Professor of Engineering, Royal Society Noreen Murray Professor in Neurobiology, Head of the Computational and Biological Learning Laboratory and Fellow of Trinity College, and was elected to a Fellowship of the Royal Society in 2012.

Making sense of data is one of the great challenges of the information age. While it is becoming easier to collect and store all kinds of data, such as personal, medical and commercial data, there are relatively few people trained in the statistical and machine learning methods required to test complex hypotheses, make predictions and otherwise create interpretable knowledge from these data. Ghahramani developed the Automatic Statistician with the aim of building a science of artificial intelligence for data. The Automatic Statistician is a system that explores an open-ended space of possible statistical models to discover explanations for a dataset, and then produces a detailed report with figures and natural-language text. Such machine learning approaches can dramatically extend access to complex statistical models so as to help anyone seeking to make sense of data.

THE PUBLIC PROMOTION OF ENGINEERING

Perhaps the most famous attack on dams was during the Second World War when bouncing bombs were used in the Ruhr Valley in 1943. Seventy years later, the Cambridge engineer Hugh Hunt was asked to act as Lead Engineer in a documentary remaking the raids. Hunt, Reader in Engineering Dynamics and Vibration and a formidable exponent of the public promotion of engineering who has won the Royal Television Society Award and the RAEng's Rooke Medal in recognition of his work, sensed an opportunity to recognise the work of British engineer Barnes Wallis and his bouncing bombs. Thus the extraordinary wartime feat was recreated for a documentary, *Dambusters: Building the Bouncing Bomb*, with Hunt leading a team of Canadian engineers and pilots. They built a 10-metre-high, 40-metres-wide dam on a lake in Canada and destroyed the dam with a third-scale bouncing bomb suspended in a rig carried by a DC4 flying just 18 metres above the surface. The documentary attracted over 10 million viewers. The fact that a single engineer had succeeded against the odds struck a chord everywhere, and hopefully a new generation of young engineers was born as they heard about the engineering challenges faced by Hunt, which included scale-model testing, design of a drop rig, targeting the dam and designing the explosive.

Decades before Hunt was born, Bertram Hopkinson had asserted that mathematics underpins engineering, and Hunt was fully aware of this when he collaborated with Guy Martin, who set a new world speed record for the Wall of Death. Does this matter? Well, nearly every schoolchild has heard of Guy Martin, and every A-level physics and mathematics student learns about 'centrifugal force'. The idea that mathematics, physics and engineering were underpinning Martin's record-breaking ride is inspirational. Afterwards, Martin claimed that he wouldn't have done the ride unless the mathematics had said that it was safe – provided he held his nerve!

CHAPTER TEN

Going West

David Cardwell (b. 1960)

Professor of Superconducting Engineering (2003–);

Head of Engineering (2014–)

David Cardwell, current Head of Department. Cardwell is Professor of Superconducting Engineering and leads the Bulk Superconductor Group.

Ann Dowling decided to step down as Head of Department after a successful five years in office and the Faculty Board appointed Sandra Dawson, Chair of the Faculty Board, to lead a Working Group with the remit of identifying a suitable successor. The 11 members of the Working Group initially drew up a list of all professors and readers, asking each of them if they were willing to be considered, as is standard modern day practice in the appointment of Head of Department. At the same time all Departmental employees were invited to give their views on these potential candidates, and students' opinions were sought through the Staff–Student Joint Committee. A shortlist was produced following this period of consultation and David Cardwell was recommended by the Working Group to be appointed for five years in the first instance.

On 1st October 2014, Cardwell inherited a Department that was vibrant and successful. Undergraduate applications were booming, with around six applicants per place, the number and quality of research students admitted to the Department was at a record high, the academic and intellectual quality of the teaching staff was second to none, the commitment and contribution of the support staff was outstanding and annual income from all sources had exceeded £70 million for the first time in the Department's history. It was also the largest department in the University, with 1,200 undergraduates, 850 postgraduates, 260 postdoctoral staff, 180 academic staff, including 56 professors and 250 support staff. But the Department was in danger of becoming a victim of its own success, as growth began to outrun the development of facilities.

Left: Aerial view of the West Cambridge site, present home of the Institute for Manufacturing, Whittle Laboratory, Electrical Engineering building, Nanoscience Centre and Schofield Centre.

Below: Artist's impression of the Civil Engineering building to be located in West Cambridge.

THE MOVE TO WEST CAMBRIDGE

A divisional structure had evolved in the Engineering Department following the retirement of John Baker in 1968, driven mainly by administrative necessity. Over the years each division had changed its research direction and grown semi-autonomously, usually with spectacular success. A natural consequence was the ever-pressing need to expand the Department, which had led to the Whittle Laboratory and the Schofield Centre being established at the West Cambridge site many years ago. More recently, two divisions had joined these facilities on the West Cambridge site: Electrical Engineering, including the Nanoscience Centre; and Manufacturing and Management.

The Department's headcount had increased at an average annual rate of 3 per cent for the seven-year period from 2006 to 2013, which had compounded the competition for resources and heightened awareness of the need for new buildings. Every division was hitting the limits of its capacity in terms of usable space, which had put a considerable strain on facilities and was beginning to stifle the potential to develop new areas of research. It was clear to Cardwell and to most of the Department that something had to change.

The separation between the Scroope House site at Trumpington Street and the new facilities on the West Cambridge site was also causing logistical and operational problems, and it was obvious to Cardwell that re-integration on a single site was the only practical way forward.

MAKING A MOMENTOUS DECISION

Before taking over as Head of Department, Cardwell had spent a sabbatical thinking seriously about his new role and the contributions that he might be able to make towards developing the Department. Cardwell wrote: 'My wife Sharon and I were hiking on Mount Cook, New Zealand's highest mountain, in January 2014. The sky was blue, the sun was shining, and there wasn't another person in sight. It was at that moment that I took the decision to move the Department to West Cambridge as my highest priority.'

Even before taking over his position as Head of Department, Cardwell began a consultation process. Fortunately for him, the urgency for creating more space had eased a little, as the new James Dyson Building would alleviate the immediate and acute problem of accommodating faculty members and students, giving breathing space and making a critical contribution towards an eventual carefully considered and well-timed solution.

The Department's International Visiting Committee (IVC), chaired by Alec Broers, former Head of Department, met on 24th January 2014, and Cardwell joined the meeting via Skype from New Zealand. Committee members listened carefully to David Cebon's analysis of the potential for redeveloping the Scroope House site at Trumpington Street and weighed this option against the

A map showing planned development in West Cambridge.

alternative proposed by Cardwell – a move to West Cambridge within a decade. The IVC, the Department's Academic Committee and Cebon reached a strong and complete consensus – the time was right for the Department to move west.

Returning to Cambridge, Cardwell decided to make the first formal public announcement of the intention to move to a meeting of some 250 assistant staff gathered in the largest lecture theatre (LT0). The assistant staff seemed to embrace the excitement of the IVC, and no major concerns were raised at the meeting. But despite the constraints of space, moving away from Trumpington Street was not acceptable to all. The arguments put forward by faculty members, postdoctoral staff, support staff and research students against moving were the great affinity for the Scroope House site at Trumpington Street and a strong preference for its central location in the city, which was considered close to the colleges and other city amenities.

However, a move west, for the whole Department, was the only option open to Cardwell. It was not a new idea by any means, as Heyman in the 1980s and each subsequent Head of Department had considered this option, and countless meetings and discussions had been held on the subject – though with no perceptible progress. Cardwell decided to make it his mission to transfer the Department in its entirety to West Cambridge by the 150th anniversary of the Department in 2025, and thus began a process of agreeing the planning principles for the design of the Department.

The planning principles addressed the development at West Cambridge, but included an essential requirement to develop and maintain the Scroope House site until the move could be completed, because otherwise the work of many staff and students could be blighted for a decade. The James Dyson Building, the Dyson Centre, the Library development, the new Fluids Laboratory and many other projects are examples of this principle being enacted.

Philip Guildford, David Cardwell and David Cebon review plans for West Cambridge.

THE JAMES DYSON BUILDING

In May 2016, James Dyson officially opened the James Dyson Building and the Dyson Centre for Engineering Design, which were built on the Scroope House site. The guests saw a wide range of student projects along with examples of the latest research and were captivated by the enthusiasm of the teams to crack hard technical problems with the intent of making the world a better place. Dyson said: 'I'm hopeful that this new space for Britain's best engineers at the University of Cambridge will catalyse great technological breakthroughs that transform how we live.' The Vice-Chancellor spoke of the ability of future generations of engineers to address the major challenges of the twenty-first century, the Head of the Engineering Department spoke of the need for collaboration for solving global engineering challenges, and Ann Dowling, as President of the Royal Academy of Engineering, said that 'Academic rigour must meet with practical invention.'

The building not only provides much-needed space but is also itself an experimental facility. Sensors have been incorporated into the building under the guidance of the Department's Centre for Smart Infrastructure and Construction (CSIC), including fibre-optic sensors in the foundations capable of measuring the strain and temperature gradients along the piles and in the concrete. Mohammed Elshafie from CSIC said that 'Installing CSIC's sensing technologies transforms the building from a passive block of material into a living creature. We will be able to ask the building how it is feeling and the building will be able to reply.' He also highlighted other features of the building, such as carbon-free electricity and the lowest energy usage across the University's estate.

PLANNING PRINCIPLES FOR THE NEW ENGINEERING DEPARTMENT

Cardwell's vision for the new home for the Engineering Department was that it must stand out as an example of good engineering practice in its design, construction and operation. Visitors from across the world should regard it as the well-engineered headquarters of Cambridge Engineering. The new facilities must meet the needs of the world's leading industrial companies, including the UK's finest, but also serve the smaller up-and-coming firms and entrepreneurial start-ups, and do so within the rigour of a high-quality academic framework. The facilities must engender close engagement with industry at all times in both research and teaching. The Departmental space should have an ambience – a 'street buzz' – with all research on permanent display and regularly updated. The building designs must portray 'engineering aesthetics', giving all occupants a sense of arrival into the central hub of Cambridge Engineering and a feeling

Above: The James Dyson Building opened in 2016. The project was made possible by a substantial donation from the James Dyson Foundation. The new building houses academics and postgraduate researchers working together across a wide range of engineering disciplines.

CSIC engineers on site during construction of the James Dyson Building.

of integration into the Department as a whole. It is essential that the design encourage staff–student interactions by creating suitable space for encounters both in the heart of the Department and along its peripheries.

A Master Plan for the new department on the West Cambridge site was subsequently developed to maximise the potential of the academic advantages of a move to West Cambridge, which include:

- the six engineering divisions in close proximity to restore coherence to the Department, which is fragmented by the current split-site accommodation
- support for the development and pursuit of the emerging strategic vision
- enabling the planned expansion of the Department over a 20–year period, which is essential if it is to remain competitive with other world-leading engineering departments
- the close proximity of key partners such as the Departments of Chemical Engineering and Biotechnology, Physics, Materials Science, and the Computer Laboratory, facilitating collaboration and the use of shared facilities

THE DYSON CENTRE FOR ENGINEERING DESIGN

In 2013, Simon Guest, in his capacity as Deputy Head of Department, identified an opportunity to develop a new design centre to support both course-related and extracurricular student-led projects on the main Scroope House site. He built on the ideas of his predecessors, Chris Morley and Richard Prager, to develop a viable plan. Guest then secured joint funding from the Higher Education Funding Council for England and the James Dyson Foundation to refurbish the space occupied by workshops on the first floor of the Inglis Building and the Department Library and build a bridge to connect them.

The project, led by Guest with the support of Hugh Shercliff, created spectacular combined space on the first floor of the Baker and Inglis buildings. The Dyson Centre embodies the spirit of the 'MakeSpace' movement that has been gaining in popularity around the UK in recent years: its facilities include an array of prototyping tools, ranging from traditional machines such as lathes and milling machines to computer-controlled machines, 3D printers and laser-cutting equipment. The workspaces are equipped with hand tools and design facilities. Richard Roebuck, Manager of the Centre, said that the students 'will come here and try to realise their ideas – try to put into practice some of the theoretical ideas they have heard in their lectures'.

People got their first glimpse of the Dyson Centre for Engineering Design as it approached completion in January 2016 with a day devoted to informal demonstrations and activities.

The dramatically enhanced space is a major contribution to the modernisation of facilities for teaching and research in the Department.

- the potential development of shared teaching, library and IT facilities with other departments, in accordance with University policy, enhancing the experience of engineering students and staff and increasing the cost effectiveness of provision for the University
- the development of state-of-the-art buildings and facilities attracting even higher calibre undergraduate and postgraduate students to the Department and increasing significantly the potential to generate income from research collaboration with international industry
- the inclusion of practical, well-planned social space for both staff and students within the infrastructure of the Engineering Department for the first time in its history
- easy access, reducing the need for many external visitors to travel through the centre of Cambridge
- a unique opportunity to establish a landmark building in Cambridge, consistent with the size and status of the Department of Engineering within the University

David Cardwell with the five preceding Heads of Department. Back row from left: David Newland, Keith Glover and Jacques Heyman. Front row from left: Alec Broers, David Cardwell and Ann Dowling.

It was clear that the construction must be planned and executed to ensure that it is necessary to dig only once. This means building at high density from the start with minimum disruption to existing buildings, yet scheduling each project to accommodate the inevitable uncertainties of fundraising. It also points towards a common, flexible platform for the progressive enhancement of services as needs arise so that upgrades can be implemented without unnecessary cost and complexity. Finally, there should be simple interfaces between separate buildings, giving flexibility in use and ease of conversion from one need to another.

PREPARING FOR THE MOVE

The University had appointed the civil engineering consulting company AECOM to prepare a Master Plan for West Cambridge, and the process was well underway without any proper provision for the Engineering Department. Just four weeks before the deadline to produce the first draft of the Master Plan, Cardwell met with AECOM together with Steve Young, member of the Department and Senior Pro-Vice-Chancellor of the University of Cambridge for Planning and Resources. To his surprise, there was a sense of relief among the planners that the firm decision had been made to move Engineering to West Cambridge, but this relief was tempered with alarm at the lack of time available for including the Engineering Department's project in the Master Plan.

AECOM arranged a series of meetings with the University and Department to include Engineering in the Plan within an additional 100,000 m² of space. David Green, Buildings Projects Manager, along with Philip Guildford, provided

AECOM with the data needed to revise the Master Plan while Cardwell worked to ensure that the Department secured the best possible site. It was eventually agreed that the Department should be situated at the far east of the West Cambridge site, enveloping the existing Whittle Laboratory, the Electrical Engineering Building and the Nanoscience Centre. This would involve extending the land for development onto the existing Rutherford and Mott buildings of the Cavendish Laboratory in the final phase of the development, once the Cavendish Laboratory had been relocated to the other side of J J Thomson Avenue, in the East Paddock of the Veterinary School.

Cardwell also appointed an Operations Committee, chaired by him but orchestrated by Cebon, consisting of representatives of all the Department's stakeholders, including divisional representatives, undergraduate and postgraduate students, assistant staff and technicians. The brief of the Committee was to look inwards, to ensure that the needs of the divisions were considered and that provision for teaching and support activities was met. The Operations Committee made it abundantly clear very early in the planning that social space should be the hub of the new building design, driven by the long-felt shortage of such facilities on the Scroope House site.

RESEARCH STRATEGY

In parallel with the work of the Operations Committee, Philip Guildford, Deputy Head for Strategy and Operations, was tasked to coordinate a substantial update of the research strategy for the Engineering Department in consultation with the academic community. This led to four themes that built on those from Dowling's strategic review:

West Cambridge will emphasise collaboration and shared spaces, creating a buzzy ambience and close engagement with industry.

- energy, transport and urban infrastructure
- manufacturing, design and materials
- bioengineering
- complex, resilient and intelligent systems.

Each theme gathered academic interests from across the existing six divisions to create communities of interest. The later addition of missions within each theme gave the necessary detail, direction and impetus for planning. Each mission addressed a significant challenge facing the world and set the agenda for how the Department could work with its international partners to help find solutions – sustainability was a strong thread running throughout this thinking, when taken to embrace not only environmental factors but also social and economic concerns. The missions connected the ambitions and interests of the academics, the strategies of the School of Technology, those of the University as a whole and those of key external sponsors. One mission was created to unify the whole plan: twenty-first century engineers. This key mission aims to inspire future generations of engineers, equip them with the best integrated engineering education, and engage them at the leading-edge of engineering thinking, so that they can change the world. Together, the themes and missions capitalise on the Department's long-held strength of being an integrated engineering department.

This structure was tentatively mapped onto the emerging layout proposed for Engineering on the West Cambridge site. There seemed to be room to fit all of the pieces of the puzzle on the land available with the right adjacencies and a good degree of flexibility – the move west looked viable. Master planning then commenced in earnest for the 100,000 m² state-of–the-art facility that could enable the Department to realise its phenomenal potential and create a thriving community of twenty-first century engineers.

ENGAGEMENT WITH INDUSTRY

Cardwell realised early on that the scale and significance of this project presented an opportunity to work much more closely with the Department's stakeholders. As a result, he decided to consult with existing collaborators, such as Boeing, BP, Dyson, Jaguar Land Rover, Maclaren, Rolls-Royce and Shell, on strategy for the new Department. Other members of the team broadened the consultation further and engaged with key partners around the world. Representatives, together with senior members of the Department including Dowling, Mair and Welland, formed the outward-looking face of the project, interfacing with industry and fundraising for the move.

In December 2014, supported by Georgina Cannon in the Development Office, Cardwell started to gather industry input to inform strategy for the new Department. The result was both unexpected and transformative: a number of clear messages emerged which fundamentally changed the priorities for the project. In particular, four requirements were formulated (see box opposite).

FUTURE PLANS AND TIMELINE
From October 2016 to 30 September 2025

The Department of Engineering has started on a great and exciting journey to re-integrate its entire activity on the West Cambridge site. The size of the challenge is enormous – the largest single departmental project in the history of the University – and the road ahead is both long and tortuous. The most challenging element is to design sufficient flexibility into the process. The build will have to be responsive to fundraising success and yet stay in line with a long-term vision. While excitement will focus on new build at West Cambridge, the Scroope House site must be maintained and adapted to provide good facilities until the very last member of staff has moved. And during the move, the world will move on and our research and teaching will evolve – the process must allow for such change.

Success will depend on careful listening and sensitivity with regard to not only staff and students in the Department, but also those in other departments.

Rolls Royce Trent 900 engine.

INDUSTRY INPUT ·

Industry would like Cambridge Engineering to:

- provide incubator space where small- to medium-sized high-tech products, close to market, could be nurtured for three to 12 months
- provide scale-up (or prototype) space where technical products at a relatively low technological readiness level could be developed
- embed enterprise and entrepreneurship more comprehensively in undergraduate and postgraduate engineering teaching
- provide industry with the potential to own premises close to the new Department.

Combined with the work of the Operations Committee, this pointed to a need for flexibility in the use of research space and for well-planned functional relationships between the physical research spaces across the divisions and research themes. It would also be necessary to accommodate visitors in a range of environments, with sensitivities concerning ownership of intellectual property, and to establish a broader facility for interacting with industry on a scale the likes of which the University had never seen before. This was the Department's opportunity to found a University-wide centre for manufacturing. It was also clear that the Department needed to be more flexible in the way it educates students, and to consider initiatives – such as a 'sandwich' year in industry during the undergraduate course, closely managed internships to augment the existing industrial experience requirements, and support for entrepreneurship and spin-out companies – to meet the changing needs of industry worldwide in the twenty-first century.

Above: Boeing's Everett Factory where many of its models, including the Dreamliner, are manufactured.

Below: Metail gives customers the tools to create 3D avatars of themselves and try on clothing virtually. The technology was created through a partnership with Roberto Cipolla's team in the Computer Vision Group.

This care must extend beyond the Department to members of the local community, local businesses and industrial partners from further afield. The plans must evolve to keep everyone on board, fully engaged and eagerly anticipating the end result.

So far, the move west project has been driven by strong commitment from the entire Department. The way academic, academic-related and support staff have given their time and energy to the move is responsible for its relatively rapid progress through the Master Plan stage, and will be critical to its success in the future.

The Department of Engineering at the University of Cambridge has an illustrious and successful past; the move west will give it the foundation for a successful future. The importance of this project goes beyond Cambridge or the UK. The world has rarely faced such a daunting array of social, economic and environmental challenges. Engineering will play a vital role in resolving many of these issues. Engineering at Cambridge is uniquely well positioned to shoulder significant responsibilities and play its role in making the world a better place for all in the twenty-first century.

List of Sponsors

This book has been made possible through the generosity of the sponsors listed below. The list has been standardised.

Philip A Abbey
Jon Abbott
John V Adams
Michael Adams
P I F Adams
Robert V Adamson
Katherine Addison
Gordon Adshead
Rajapillai Veluppillai Ahilan
Mahmood Ahmed
Frank Ainger
Mark Ainslie
James E Aitken
Marwa Al-Ansary
Abir Al-Tabbaa
Lawrence Albinson
Michael Albutt
Nicholas Aleksander
Rami Alghamri
Duncan Allen
Nigel M Allinson
Patton M Allison
Martin Allnutt
Andrew Allsop
Jimena Alvarez and
    Cristian Barrios
Nadun Alwis
Theocharis Amanatidis
Gehan Amaratunga
Christopher Ambler
Joy Anderson (née Pennell)
Richard L Andersson
Colin Andrew
Keith Andrews
Jannis Angelis
Chris Angus
Roderick Annandale
John A Anning
Joseph Antebi
John Anthony
Jolyon Antill
Philip M Antrobus

Paul Appleby
Kelvin Appleton
Duncan Apthorp
Kumar Arasu
Colin Armstrong
Martin Armstrong
Johannes C Arning
Olav Arnold
John F C Arthur
Zoe Arthur-Kenealy
Freddie Ashford
John Ashton
Philip Ashworth
David Aspinall
Eric Atherton
David Atkinson
Simon and Jowan Atkinson
Simon Reay Atkinson
Martin Attfield
Rachel Attwood
Matt Atwood
Simon Au
Julian Aucken
Daniel J Auger
Duncan Auterson
David Axford
Nigel Ayton
Azmi Bin Azeman
Julie Bacon
Michael Bacon
Nigel Bacon
Aaron Baillieu
William S Bainbridge
Stuart G Baines
Chris Baker
Grahame Baker
Roger Baker
Andrew Baldwin
Bill Ball
David W D Ball
Ken Ball
Kenneth Ball

N R Ball
John Ballard
J S Banks
Donald Bannister
Daniele Baraldi
Rodney G Barber
Chris Bardwell-Jones
Laurence Barker
Claire Barlow
Michael Barlow
Mark C S Barnard
William Barnard
George D Barnes
Simon A Barnett
Francis Baron
Ian Barrett
Barry Barton
Yvonne Barton
Ajit Basak
Dali Basak
Shreeja Basak
Murat Nihat Basarir
Bill Bashforth
Bob Bastow
Ernest M Bate
Herbert Bate
Jolyon Bates
George Battrick
John Baylis
David Beadman
David Beare
Peter Bearman
Huw Beasley
Tom Beaumont
Chris Begg
Andrew C Bell
Peter J Bell
Chris Beney
Peter Bennell
James Bennion-Pedley
Robert Bentall
Edward Bentley

Roger Berkeley
Emil Bernal
Charles Betts
Michael Bevan
Mihir Bhushan
David Bickley
Tim Bilham
Michael Billington
Ashley Birkbeck
Adrian Birt
Tony Bishop
Paul Blackborow
Peter Blair-Fish
Claire Blakeway
David Bland
Peter Blatchford
Kathryn Blencoe
Thomas P Bligh
Tony Bocock
Stephen Bogod
Aleksandr Bogomil
Scott Boham
Thomas Bohné
Nicholas Bolton
Andrew Bond
Peter W Boorman
Andy Booth
Edmund Booth
Rockley Boothroyd
Nick Borner
Sarah and Roger Bostock
Charles Botsford
Chris Botsford
Charles Boulton
Paul Bounsall-Hughes
Dave Bowen
Robert Bowles
David Boxford
Grahame Boyes
Daniel Brackenbury
Chris Bradfield
Arthur J Bradley

Mike Bradley
Michael Bradnam
Tony Bramley
Graham Brand
Markus Brandstetter
P L Bransby
Beth and Kelvin Bray
Miloš Brčkalo
Peter Bream
Emma Breese
David Brereton
Bernie Breton
Mihai Brezeanu
P Bricknell
J Ramsay Brierley
Gregory Briffa
Christopher G Briggs
Peter Bright
Evangelia Brilaki
Stuart Briscoe
John Bromell
Michael Brookbank
Edmund Brookes
Steve Brooks
A R (Tony) Broscomb
Alison Brown
Andrew Mark Brown
Ele Brown
Martyn C Brown
Phil Brown
Richard H Brown
Steve Brown
Walter Brown
Steve Brumpton
Elisabeth Brun Lacey
Angus Bryant
Geoff Buck
Steve Buckley
Ian Bucklow
Graham Budd
Dave Budenberg
Adrian Bull

Graham Bull
Barney Burgess
Colin Burgess
Richard Burkinshaw
John B Burland
Bruce Burnett
Paul Burnham
Graham Burr
Francis M Burrows
F J W Bush
Stephen Bush
Bill Butcher
Hans Butler
James Butler
Colin Butters
Bill Byrne
Francisco J Cabadas Rodriguez
John Cain
David Cairns
Ian Calderbank Downing
Peter Calderbank
Nicholas Caldwell
C J Caldwell-Nichols
Chris Calladine
Amanda Calvert
Cambridge Design Partnership
Ian W Campbell
Roger Camrass
Steve Canadine
Audrey Canning
Howard Canning
C Cannon-Brookes
David Cardwell
Peter Martin Carew
Kevin Carleton-Reeves
Win Carnall
Angus Rennie Carrick
John Carroll
Brian Carter
Mervyn D Carter
Paul Carter
Martin Casey
R J T Casinader
Ian Castro
Aaron Chadha
Martin Chalons-Browne
Tao Soon Cham
Jonathon Chambers
Mervyn Chambers
Catherine Ka Man Chan
Fiona Chan
Ignatius Chan
Nigel Chan
Willy H W Chan
C Richard Chaplin
E J C Chapman

Alistair Chappelle
Demetris Charalambous
Selwyn Charles-Jones
Howard Chase
Jayanta Chaudhuri
Talay M Cheema
Cheltenham College
Chih-Chun Chen
Jingtao Chen
Johnson Chen
Liu Chengxiang
Andrew Chenhall
Michael Chesshire
Royce Chew
Siau Chen Chian
Nav Childs
Andreas Chirou
Richard Chisnall
Sze Ning Chng
Rebecca Cho
Hyunmook Choi
Akiko and Ellis Chong
Tiffany Chow
Martin Chown
Richard Christmas
Daping Chu
Tze Meng Chua
Joon Huang Chuah
Tomás Chubb
Richard C W Church
Nicholas Barry Menzies Clack
Robert W Clark
Simon Clark
Christopher J L Clarke
Peter M C Clarke
Ross Clarke
John Clarkson
Dick Clements
Simon Clephan
John Jay Climenhaga
Peter Clutterbuck
Douglas Clyde
Robert Cochrane
Christopher Cockcroft
Steven N Cocking
Bernard Coffey
Adam Cohen
David J Cole
Matthew T Cole
Charlotte Coles
David Coles
James Richard Colgate
Robin Collet
Nick Collings
Andrew Collins
Janet Collyer

Derek Colman
George J L Coltart
Michael Colyer
Andrew S Connolly
Robert F Conti
Adrian G Cook
John Cook
R F Cooke
Roger Coombes
B A Cooper
Nicholas J P Cooper
Peter Cork
Marco Costanzi
Barney Court
James Cousen
G H R Couzens
Alan Cox
Tony Cox
Paul Cozens
Mark Crabtree
F J P Crampton
Jack Cribb
Simon Croft
Philip M Cross
Peter A Crossley
Steve Crowther
David Cruttenden
Nigel J Cubitt
John Cuckson
The Cullen family
Alison Cullen
Roy Cummings
Joseph P Curtis
Denys Cussins
Julian Dale
C John Danby
Graham Darke
Mark S Daskin
Tristan R G Davenne
Phillip R Davenport
Chris Davey
J F Davidson
Bob Davies
Ciaran Davies
D W N Davies
Gary Davies
Hannah Davies
Hywel Davies
Martin Davies
R W Ll Davies
Simon R Davies
John D Davis
Trevor Davis
Nicholas Davison
Michael R Davy
Peter H Dawe

Norman W F Dawson
Peter Dawson
John Day
John de Figueiredo
Vincent de Gaultier de
   Laguionie
Matthijs de Kempenaer
J R Deane
Matthew DeJong
Domenico del Re
Olusomi Delano
Philip Denbigh
Christian Deverall
Toby Dickens
Jonathan Dicker
Michael Dickson
Howard Dilley
Konstantinos Dimitriou
Jack Dinsdale
Keith Distin
Malcolm Dixon
David Dobson
James Ian Dodds
R M Doe
Tim Doggart
Phil Doherty
Alan Dorman
John Dowell
Ann Dowling
Geoffrey Drake

John Drewery
Brian Drewitt
Peter Dring
Jonathan M Duck
Michael F Dumont
Gavin C M Dunbar
John Dunham
David Durst
Ronald A Dutton
S Dwarakanath
Paul Dyke
David J Eagle
William H Earle
David Eaves
Keith Ebden
Don Eccleshall
Alastair Eddie
Vincent Edkins
Finlay Edridge
J A S Edwards
Charlie Efford
Philipp G Egger
Keith and Charles Eickhoff
Abdullah Mohammed El
   Sayed
Peter Eldred
Charles F Elias
Michael Elliott
Richard J I Elliott
Alun T Ellis

Leslie Elliss
Stuart Elmes
Nathan Elstub
Peter English
Chris J Evans
Colin Evans
Duncan Evans and
    Elizabeth Collins
Hugh Evans
Martin Evans
Mike Evans
Robert L Evans
Roddy Evans
Doug Everard
Paul Everard
Thomas Everhart
David Eyton
James Eyton
C Paul Fairweather
Roy Farmer
Guy Farnsworth
Paddy Farrell
David Faulkner
John Feavearyear
Pamela Fennell
Richard Fenner
Martyn Fice
Andrew Figgures
Matthew Fitton
Stephen Fitzgerald
Norman Fleck
Robin Fletcher
Peter Flinn
Andrew Fogg
H William Fogle Jr
Forbes Mellon Library,
    Clare College
Jack Forrester
Ian Forster

John Charles Foy
Jayné E Franck
Robert Frayling-Cork
Robert Friedlander
Marc Fry
Gareth Funk
Marcos Massao Futai
Ranjit Galappatti
Andrew Gales
Mark J F Gales
Phillip Gales
Charles H Gallagher
Lian Gan
Dennis Ganendra
Jessie Lu Gao
Bill Gardiner
Kenneth Gardner
R J M Gardner
Henriette Garmatter
Mark Garner
Neil Garrard
Nick Garrett
Martin Garrod
Bill Garwood
Patrick Gaskell-Taylor
Timothy D Gasser
John Gatiss
Michael Gavin
Martha Geiger
Martin Geiger
Kate Gentles
David George
Martin H George
Antonis Georgiou
Robin Gerard-Pearse
Petros Giannaros
Ioannis Giannopoulos
Noel Gibbard
Keith Gibbeson

Declan Gibbons
David Gibbs
R K Gibson
Reginald J Gibson
Alastair Gilchrist
Trevor Gill
Guy Gillett
Charles Gillman
Michael James Gleeson
Brian Glover
Keith Glover
Robin Porter Goff
Hamish Goldie-Scot
Terry Gooch
Richard Goodhead
Colin Goodman
Alastair Goodson
Arjun Gopalakrishnan
John Gosden
Dimitri Goulandris
Nigel J D Graham
Nicholas Granger-Brown
Andrew J Grant
Colin Grant
Andrew H Gray
Robin C Gray
Christopher Green
David Green
John E Green
Mark Green
Richard Green
Stefan Green
Tom Green
Rob Gregory
Malcolm Greig
Marr Grieve
Nigel Griffiths
Richard Grisenthwaite
Keith Groves

Chang Shin Gue
Chang Ye Gue
Simon Guest
Edward Gummow
Amit K Gupta
Chris Guy
Dave Gwilt
Ahmad Habibian
Oliver Hadeler
Paul Hagger
David C Haigh
Stuart Haigh
Crispin Hales
David Halfpenny
Christopher Hall
David Hall
John D P Hall
Peter Hall
R Hall
Richard Hall
Stephen J Hall
Neil and Lydia Halliday
Roy Hamans
W A H Hamilton
John Jo Hammill

Lou and Pat Hammond
Paul A Hammond
Robert Hampton
Gordon Hannah
Oliver Harding
Peter Hardyman
George Hare
J Derek Harling
Paul Harmer
Kenneth H Harper
L Roy F Harris
Simon Harris
David Harrison
Geoffrey Harrison
Michael E Harrison
Tim Harrison
Jonathan Harry
Giles Harvey
John Harvey
M H Hasan
Jim Hase
D G Hasko
Lindsey Haslegrave
Matthew Haslett
Maria Jimena Hatchman

Dimitrios Hatzis
Keith Haviland
Adrian Hawkes
Nick Hawkins
Guy Haworth
Joanna Hawthorne Amick
A W R Hawthorne
Elizabeth Hawthorne
    (O'Beirne-Ranelagh)
Malcolm Hawthorne
Kiyotaka Hayakawa
Edward W Haynes
Jason N Haynes
John Richard Haynes
Carl Haythornthwaite
John Hazlewood
Jeremy Hazzledine
Haitao He
Qiurui He
Ken Head
J R W Hearn
Sheryl Heath (née Scarisbrick)
N M F Heaton
G and N S Heavyside
Nick Hedges
Barry Hedley
Thomas Heesom
Mark Hempson
John Michael Henderson
Ross Henrywood
Nigel J Hepworth
Steven Herbert
M Hernu
Philip Hewer
Ewan C Hewitt
Julian Hewitt-Smith
Christopher Hewlett
Terry Heymann
Harry J Hibberd
A D Hibbs
S A Hibbs
Alan Hickling
S J Hicks
John Higgins
Peter Hilborne
B R Hill
Stuart Hill
Rod Hine
Peter Hirst
Peter Ho
Stephen D Hoath
Andrew Hobbs
Caroline Hobbs
Mark Hoffman
Stephan Hofmann
Olivier Hofstede

David Holburn
Harriet Holden and
    Christopher Pedersen
Maria Holgado Granados
Donald Holliday
William A Holloway
Nicole Holmes (née
    Humphries)
Mark Holst
David Hook
Andrew Hoole
George Hopes
Pippa Hopkins
Ingrid Hopley
Roger Horrell
Tim Houghton
Neil A Hoult
Colin Housby-Smith
Ann and Sean House
Greg Howard
Bill Howgego
Archie Howie
Michael Hoyle
Yinxiao Huang
Alastair Hughes
James M I Hughes
M A Hughes
Antony M Hulme
Mark Humphries
Brian Hunt
Hugh Hunt
Paul E Hunter
David Hutchinson
George W Hutchinson
Ian Hutchinson
Richard A Hyde
W Rob Ibberson
Daniel Imhof
Edd Inglis
Pygmalion Ioannides
Waseem M Iqbal
Peter Irvin
Michael Isaacson
Charles Isitt
Srinath R Iyengar
David Jackson
Peter Jackson
Peter Jackson
Laura James
Richard D James
Ronald James
Amy and Waldemar W F Jap
Paul Jarman
Sandy Jayaraj
Peter Jeanneret
Robert F Jeays

Stephen L Jeffels
John Jefferis
Michael Jeffery
David Jenkins
Peter Jenkins
Alwyn Jennings-Bramly
James Jennings
Stephen Jennings
Richard Jenyns
L O F Jeromin
Fei Jin
Anthony John
Andrew Johnson
Barry Johnson
Colin Johnson
David Johnson
G Keith Johnston
Innes Johnston
Colin Jones
G S Jones
Gavin Jones
Glyn Jones
Glyn Jones
J S E Jones
Leonard L Jones
Timothy A Jones
Victoria Elizabeth Maria Jones
Anthony Jordan
Graham and Jean Jordan
Oliver Jordan
Tony Judd
Chris Juden
Mohd Shafie Jumaat
Jeffrey A Jupp
Henry D Kafeman
Christos Kakoutas
Derek Kalyanvala
Chris Kamara
Antonios Kanellopoulos
Riki Kangwai
J Richard Kay
Mike Kearney
Michael J Kelly
William R Kemp
Andrew William Kenchington
Iain Kennedy
Clive Kerr
Gordon Kerr
Eric Kerrigan
Dan Kershaw
Michael Kershaw
Mohammed N Khalil
Rafil Khatib
Hassan Khawaja
Kit Soon Khong
Johan Khoo

Giles Kilbourn
Christopher King
Jason M King
Lionel C King
Nick Kingsbury
Roger Kinns
Timoleon Kipouros
Michael A B Kirk
Marwan Kishek
Karsten Klein
Sorano Franziska Klein
Ben Knightly
Reginald Kogbara
F K Kong
Valencia Joyner Koomson
Uwe Kopke
Juha Korhonen
Zbigniew Kowszun
Neil Kraewinkels
Michael Kraftman
Chris Kraushar
Hans Kühn
Praveen Kumar
Nicholas S C Kwok
Geoffrey Lack
Peter H Ladbrooke
Geoffrey Lafford
Matthew J Laight
Sandy Laird
Andrew Lait
Derek Ian Lake
George F Lake
Simon Lamb
Edward B Lambourne
David Lancaster
John Lane
Charles Langler
Hugo Larsen
D W M Latimer
Martin Lau
Philip Fat Kit Lau
Fiona M P Lauder
Meron Lavy-Moheban
David Lawrence
Bob Lawson
C S Lawson
Michael Le Flufy
Martin Leather
Chung Shek Lee
Ian R Lee
Ming Kai Lee
Nicholas Lee
S H Lee
Tricia Sook Ling Lee
Mark Leeson
Jochen L Leidner

Andrew Lennon
D C Lennon
Emily Lester
A W C Leung
Bonnie H Y Leung
Johnson H K Leung
Jake Levi
Peter Levi
Benjamin Levine
Syn Pau Lew
Alwyn Lewis
Chris Lewis
Mark Lewis
Michael Lewis
Grant Lewison
Anthony Ley
Alan Thomas Leyland
San Tat Li Kim Mui
Benjamin Li
Bo Li
Jessy Tong Li
Jian Li
Kun Li
Jiaming Liang
Chia Hui Lim
Chwee Teck Lim
Sze-Xian Lim
David Lindley
Joel Lindop
David Lines
Trevor Linnecar
P M Lister
Chrysoula Litina
Guy Littler-Jones
Bingqian Liu
Feiyang Liu
B J Livingston
Thomas Livingstone-
    Learmonth
Adam Lloyd
Jessi Lloyd
Martin Lockett
Colin Lockie
Susan Long
Teng Long
Tim Longstaff
James Losh
A W Loten
J Brian Lott
Andrew Love
Angus Low
H Y Low
Judith Lowe
Richard Lowenthal
Guoxing Lu
Jiahui Lu

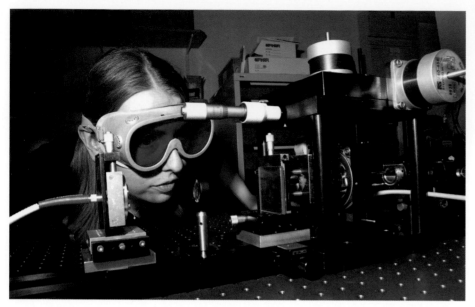

Tianxin Lu
Christopher C K Lucas
Rex Lucas
John Lucia
Jacky Lui
Sam Luke
John Lumley
Rod Lynch
Jim Lyon
Philip H M'Caw
E J Y L Ma
Fang Ma
Yunfei Ma
James MacDowell
Alistair Macfarlane
Jan Maciejowski
Neil MacInnes
Kenneth Mackenzie
Ian Mackley
Malcolm Macleod
Gopal Santana Phani
  Madabhushi
Yiling Mah
Anja M Maier
Richard Mair
Robert Mair
Frederick F C Mak
Kim Mak
I Manners
Rhoda M Manook
John Mansfield
Wenting Mao
Gordon Hamilton March
Dominic Marnell

Martin Marriott
Harry Marsh
John Marsh
Peter Marsh
Christopher Jamie Marshall
David Marshall
John Marshall
Alastair McLeod Martin
James Martin
Leopold Martin
Peter Mash
Mohamed Mashaal
Ahmed Mashhour
Guy Mason
Peter Mason
James Matheson
David Matthewman
P D Matthewman
Michael Gordon Matthews
T A M Maula
Edward Maunder
Robin Maxwell
Tim Mayne
Ian McAlpine
Strachan McDonald
Christopher McDouall
Duncan McFarlane
Andrew McGill
Gerard McGlew
Christine McHugh
Alastair McKay
Roger McKenning
Beverley J McKeon
Ian McLaren

Gordon B McMillan
Neville McMillan
Hilary McOwat
Jim McQuaid
Gordon McQuire
Jeremy J Meek
Roger Meli
Clive Mellor
Ioannis Menikou
B L Merrett
Simon Middleton
Simon Middleton
J D Midgley
J T J Midgley
Vincent J Mifsud
James Miller
Jonathan I Miller
Malcolm Miller
Benjamin Mills
Stuart Mills
Ian Milne
Olivier Ming
Tim Minshall
Inder Mirchandani
Darren Andrew Mitchell
David Mitchell
Ian Mitchell
Peter Mitchell
Stuart Mitchell
Chadwick Mok
Charles Monck
Tony Monk
Charles M Monroe
Nigel Montagu

Francisco A T B N Monteiro
Kenneth B C Montgomerie
John Moore
Christopher L Morfey
Tim Morgan
R D Morgan-Smith
Guido Morgenthal
Charlotte Morley
Alastair Morris
David Morris
Philip Morris
Raymond Morris
Colin Morsley
Nicholas Morton
David Moss
Jeremy Moss
Rachel E M Moss
Jairo H Moyano
Philip and Catherine Mulcahy
Brian E Mulhall
Fadi R Muna
Mohan Munasinghe
Murray Munro
R G Munro
Simon Munton
Peter Murfitt
Paul Musgrove
Alan Mushing
Denis Mustafa

James Mynors
Cristian Nacht
Purnendu Nath
Bruce Nathan
Andy Neale
Paul and Maureen Negus
John Neilson
David Neish
Preben Monteiro Ness
D E Newland
Peter E Newley
Eddie Yin-Kwee Ng
Erin Ng
Wai-Yin Ng
Bella Nguyen
Charles C Ni
Richard J R Nicholas
Mark H Nicholson
David Nickols
Tony Nickson
Doros Nicolaides
Ajit Nimalasuriya
Eni G Njoku
Richard A D Noble
Keith Norman
Norton House (HK)
John Nutt
Dermot O'Callaghan
Michael O'Connor

Jerome O'Hea	Dunstan Pedropillai	Michael Purshouse	Nigel Robinson	Luke A Scott
Rachel Oakley	Chris Pemberton	Michael Pye	Patrick Robinson	F P Scourse
John Oatley	Martin Penney	Simon Pyzer	Paul Robison	Jack Seaborn
John Ochsendorf	Alex Petrie	Michael Qian	Cuauhtemoc Rodriguez	Videsh Seereeram
David Odling	Andrew Petrie	Christopher Quarton	Geoff Roe	Robert G M Selby
Brian Oeppen	Hugh F Pettifer	David Quarton	Roger Rooke	Aditi Sengupta
Leonard D Ogden	Adam Phillips	John Race	John Roper	Ashwin A Seshia
Keith Oldham	Matthew Phillips	Andrew Rackham	'Roger' Roper	Martin Sessions
Bobby Foo Yew Onn	Brian Phillipson	Hedley Railton	Nigel Rose	Craig Seymour
Nigel Orchard	Neil Pidduck	Sivasubramaniam Ramanan	Grahame Rosolen	Peter Sharp
Tony Orchard	Simon Piggott	Helen V M Ramsay	John S H Ross	John Sharpe
Claudiane Ouellet-Plamondon	William C Pike	Munidasa Ranaweera	Michael M B Ross	Malcolm Sharples
Andrew Overy	Simon Pilgrim	Peter David Ransome	Tim Rossiter	Tim Sharrock
Anne Owen	Chris Pilich	Tankasala Ranga Narasinga Rao	Kenneth Rotter	Christopher T Shaw
Philip Clifford Owen	John Pilling	Robert Rattee	Arthur Rowe	G B H Sheah
Vassilis P Pachiyannis	Shusanah Pillinger	Ben Ravenscroft	Peter Rowsell	Leila Shelley
Richard Paden	Neil Pinto	Robert Rawlings	Erik Rudzitis	Peter Shelswell
Frederick B M Page	Ben Piper	Steve Rawnsley	Fu Ruizhi	Xiangyu Sheng
Stephen Page	Dorothy Pipet	Keith Ray	Chris Rupp	Mark R H Sheppard
Alastair Paice	Henry Ian Pizer	Bryan C Read	E D Ruppel	Hugh and Tessa Shercliff
Paul Palfreyman	Jim Platts	Jack Read	D F H Rushton	Richard Sherratt
Andrew C Palmer	Judith Plummer Braeckman	John Read	Peter H Ryan	Nigel Sherwood
Clifford Palmer	Stephen Plummer	John Read	Leslie Rysdale	Peter L Sibley
Mike Palmer	David Pocock	John Reddaway	Martin Sachs	Richard B Siddall
Sebastian Palmer	Stuart Pomeroy	Sadhana Reddy	Tunde Salami	Miguel Sierra
Stephen Palmer	Vaughan Pomeroy	Peter Redfern	Eric Hanson Sale	Anthony Silas
Robert Palmström	Andy Pomfret	Pat Redlich	John Richard Hanson Sale	Paulo Silva
Darius Panahy	Paul Poncelet	Clive Rees	T C H Sale	Avi Sinharay
Aron Ki Wai Pang	Bryan Man Hay Pong	David Hugh Rees	Ramy Salemdeeb	K Z Sirwan
Richard Pannett	Benjamin Poole	Jeremy Rees	David Saltmarsh	Anthony Skaife d'Ingerthorpe
Ares Papangelou	Jeremy Poole	Paul C Reid	Imantha Samaranayake	Christopher Sketchley
Priti Parikh	Henry Poon	Seena Rejal	Asela Samaratunga	Howard Skipp
Philippa Park	Richard Popplestone	Rory Reynolds	Ferdinando Samaria	Charles Slater
Shaun Parker	Jason Leith Porter	Rasha Rezk	Graham Sampson	Roy Smalley
Christopher Parkes	Henry H Potter	Kyriakos Rialas	Eskandarian Samsudin	Mark Smallwood
Douglas Parkes	John O Pounder	Sarah Rice	Rodrigo Sanchez-Mortensen	Derek R Smith
Geoff Parks	Jolie Powell (née Carter)	Peter Richards	Caroline Sanders	Graham D J Smith
Toby Parnell	Nicki Power (née Rodale)	Robert J Richards	Roby Ferreira Dos Santos	Ian Smith
Tom Parrott	Richard Prager	Rachel and Nigel Richardson	David Sapiro	Ian F C Smith
John Parsons	Patrick R Prenter	Jim Rickard	Jeremy Sargent	J Kent Smith
Anil Patel	Adam Preston	Paul Riglar	Yoshiaki Sasamura	Jonathan Hurndall Smith
Graham Patey	Anthony Price	Keith Riley	H Martin Saunders	Luke T W Smith
Donald Paton	Emma-Jane Price	M L Rivaud-Pearce	Jeremy Saunders	Malcolm Smith
Nigel Patrick	Jonathan Price	Elwyn Roberts	John Savage	Michael Anthony Smith
Phil Patrick	Martyn John Pritchard	Gwilym M Roberts	Mark Savill	Mike Smith
Michael J Patton	Doug Probert	Peter Roberts	Seb Savory	Paul Smith
James Paul	Michael J Provost	Bruce Robertson	Dhruv Manmohan Sawhney	Roderick A Smith
Andrew Pauza	Ian Prowse	Douglas Robertson	Matthew Scarisbrick	Stephen J B Smith
Roger W Payne	Jonathan Pryke	Ian Robertson	Paul Scarisbrick	Stephen L Smith
Simon Payne	Peter Melburn Pucill	John Robertson	Raphael Schmetterling	Steve Smith
Christopher Payton	H J H Pugh	John Robertson	Walter Schroeder	Christopher Smyth
Joseph Payton	Penny and Malcolm Pullan	Morven Robertson	Matt Schumann	Peter Solomon
John M Peake	David Pullman	A J M Robinson	Kurt Schwarz	Phil Sorrell
Katherine and Steven Pearse	Anthony Purnell	Charles Robinson	Hugo Scott Whittle	Constantinos Soutis
D M Pearson	W Pursell	David Robinson	J Nigel D Scott	Donald Spiers
J R A Pearson	Richard John Michael Purser	J A Robinson	John Scott	Timothy W Squire

Bryan Stead
Walter R Stead
Chris Stenton
P R Sterland
Leonard Stevens
Robbie Stevens
James Stewart
Malcolm Stewart
Geoffrey Stickland
J Anthony Stiles
Alastair Stirling
Trevor Stirling
Andrew and Henrietta Stock
Laurence P Stock
Nigel A Stoke
Michael Stone
Neil Stothard
Richard Stradling
Oliver Stratton
Steve Street
Martyn Stroud
Jonathan Stuart
Cathy Stubbs
Peter Stuckey
Jonathan and Carol Such
Philip L Sulley
Ananda Sumanadasa
Peter T Summers
Victor Sun
Stephen Sunderland
John Sutton
Peter Swales
N Swaminathan
Robert Swann
Charles Sweeney
Alan Swindells
Neil Sykes
R John Sykes
Graham Talbot
James Talbot
Huei Ming Tan
Lin Tze Tan
David A Tanqueray
Daniele Tartarini
Ben Tatham
David J Taylor
Donald Taylor
J Graham Taylor
John Taylor
John Taylor
Paul H Taylor
Peter Taylor
Peter Taylor
Phil Taylor
Richard Taylor
Robert Taylor

Juliet Teather
George Tedbury
Hilary Temple
Joseph Tevaarwerk
Berrak Teymur
Dhama Thanigasapapathy
N Thedchanamoorthy
Robert G Thomas
Ian Thompson
J Michael T Thompson
Richard H Thompson
John Thorn
Michael F C Thorn
Roger Thornber
Alan Thorne
Richard Thorne
Roland Thorp
Christopher Tidy
Frank Tietze
Steve Tilsley
Paul Timans
Nicholas Timmins
Peter Tinsley
Neil A Titchener
Philip Titheridge
Richard Wojciech Tobiasiewicz
Ben Todd
C M Tomkinson
Nevil Tomlinson
Sarah Shin Yee Tong
Jason Too
Ken Totton
Keith Tovey
Martin Tovey
Anthony Townsend
David Tron
Michael G Trotter
Matthew Trowbridge
Christopher Trye
H K Tsang
Mark Chau Shing Tse
T S Tse
King-Jet Tseng
Naoum Tsolakis
Bertha Tsui
Tee Boon Tuan
Julian Tuccillo
Keith Turnbull
David A Turner
Iain Turner
John Turner
Peter Turner
Steve Turner
Kit Twigge-Molecey
Samuel Twist
Florin Udrea

Chloe Underdown
Michael Underwood
Jack Upsall
William Usher
Shunichi Usui
Peter van de Kasteele
John Vandore
T Paul J Vardanega
Savio S V Vianna
John Vignoles
M T Vijay Vijayendran
Matthias Volmer
Peter H Von Lany
John Waddington
Richard Wade
J W Wadkin
Adrian L Waghorn
Shu-Kee Wai
Peter Wake
Ian Wakeford
Mark Wakeford
Robert Wakeford
A F Walker
Adrian Walker
D L Walker
Francis H D Walker
Ian Walker
Miles Walker
Robert David Walker
Tim Walker
Ken Wallace
David Wallis
Tony Walters
Hainan Wang
Hao Wang
Hua Wang
Shangyi Wang
Suhan Wang
Sumeng Wang
Yalou Wang
James Ward
Roger Ward
Scott Warden
Dick Warwick
Kamol Wasapinyokul
A B Wassell
James Waterton
A J R Watson
David Watson
Mike Watson
Rob Watson
Andrew Watts
Thomas Watts
Lance Waumsley
David J Way
Raymond Weatherby

Paul Webb
Claire Weiller
David Weiss
Colin Wells
Howard Wells
Stephanie M C Wen
Yu Sheng Wen
Martin West
Charles Westgarth
Michael James Weston
Peter F Whatling
Adrian Wheal
David Whitaker
Andrew White
David White
David White
Ian White
Donald Whitehouse
Steven Whitehouse
David Whitfield
K R Whittington
Richard M Wightwick
Ray Wijewardene
Tim Wilkinson
Patrick Willan
David Willans
David Williams
Fred Williams
Jeremy and Alexandra
    Williams (née Bryans)
John A Williams
Roger D Williams
Vivian P Williams
Ted Willmott
Brian Howard Kelway
    Willoughby
Adam Wilson
D R Wilson
Gerald Wilson
James Wilson
Keith R Wilson
Piers Wilson
Richard Wilson
Tim and Margaret
    Winchcomb
Mike Windle
Jon Windsor
R Wingate-Hill
David Winter
Mike Winterbotham
Nicholas Withers
Daniel Wolpert
Ford-Long Wong
Gordon Chung Leung Wong
Hiu Hei Wong
K K Wong

Kai Jiun Wong
Terence K S Wong
Adam Wood
David Muir Wood
M Wood
Mike Wood
Charles Woodburn
Peter Woodburn
Nigel Woodford
Lindsay Woodhead
Andy Woodland
Alan Woodley
John Woodman
John F Woodward
David Worlidge
John Worlidge
Bruce Worrall
Christopher Worsley
Stephen Wright
Stuart Wright
Qian Wu
T A Wu
David Wyatt
Geoffrey Wyss
Yongsong Xie
Changwei Xu
Haofeng Xu
Haobai Xue
Anson Yan
Fanjun Yang
Shu Yang
Zhongliang Yang
James Yea
Jeffrey Yee
Ping-Yi Yee
Krishnaswami Yegnanarayan
Jim Yelland
George Yeo
Mozhi Yin
Alfred Yip
David J Young
Jenny Young
John Young
Steve Young
Tongxi Yu
Shaowu Yuchi
Akhmad Herman Yuwono
Frank Zachariasse
Nadine Smadi Zakaria
Steve Zan
Qi Zhang
Yiyun Zhang
Yunhui Zhang
Yaoyao Zheng
Victor Yiqian Zhuo
Krzysztof Zuber

Index of Names

Picture Credits

Many of the images in the book are from the Department of Engineering's archive. The Department and Third Millennium would like to thank Alan Davidson of Stills Photography for taking photographs specifically for this publication, Haroon Ahmed and Peter Long for providing and giving advice on image selection, Christopher Jablonski for his hard work in tracking down images, as well as the individuals and organisations listed below for kindly giving permission to reproduce material. In the case of an inadvertent omission, please contact the Publisher.

6, 7, 8–9, 9, 11bl, 11br, 24, 26–27t, 28b, 35tl, 35br, 36–37, 43tl, 43b, 44–45, 45t, 46, 47b, 48, 51, 53, 55tl, 58–59b, 59, 61m, 61b, 62t, 66, 68, 68–69, 69, 70, 71, 77r, 77b, 78, 80–81, 82–83, 84, 84–85, 86, 87, 88–89, 89, 90, 92, 92–93t, 95, 96t, 96–97, 97, 98, 100, 101, 102–103, 104tl, 104tr, 105b, 106–107, 108–109, 111tr, 111r, 111br, 117, 118, 130tl, 136b, 147, 160t, 180, 186 Department of Engineering archive; 2–3, 4–5, 8, 10, 11b, 18t, 22, 23br, 29, 30t, 31r, 34, 35tl, 36–37, 39r, 40–41, 42, 43tr, 47t, 50, 54, 55tr, 58–59t, 68, 76, 91t, 102, 103, 110, 112–113, 115t, 116t, 119b, 122t, 124t, 124–125, 127, 129, 135m, 140–141, 141, 145b, 146b, 148, 149, 157, 168, 170, 173, 175, 179, 185, 188, 191, 202br Stills Photography; 11tr Michael Juno/Alamy Stock Photo; 11tl Private Collection/© Look and Learn/Valerie Jackson Harris Collection/Bridgeman Images; 12t H S Photos/Alamy Stock Photo; 12b Courtesy of the President and Fellows Queens' College Cambridge; 13, 33 Science & Society Picture Library/Contributor/Getty; 14 Print Collector/Contributor/Getty; 15 Fitzwilliam Museum, University of Cambridge, UK/Bridgeman Images; 16 © National Portrait Gallery, London; 17 The Master and Fellows of Trinity College Cambridge; 18b National Galleries of Scotland/Contributor/Getty; 19 Chronicle/ Alamy Stock Photo; 20–21 Private Collection/The Stapleton Collection/Bridgeman Images; 23tl Jim Gibson/Alamy Stock Photo; 23tr Dave Hollos; 23bl University of Dundee Archive Services; 24–25 General Library of the University of Tokyo; 25 Wellcome Library, London. Henry Charles Fleeming Jenkin. Etching by W Holl, 1884; 27br Dallas Museum of Art, Texas, USA/Junior League Print Fund/Bridgeman Images; 28t Photo by SSPL/Getty Images; 32 © Science Museum/Science & Society Picture Library; 32–33 Royal Collection Trust © Her Majesty Queen Elizabeth II, 2016/ Bridgeman Images; 35tr from *Engineers and Enterprise: The Life and Work of Sir Harry Ricardo* by John Reynolds, Sutton Publishing. Published in 1999; 49 'Metamorphosis of the airplane over the period 1918–44' from a paper by William S Farren to the Institute of Aeronautical Sciences New York, 1944; 52t The Master, Fellows and Scholars of Downing College in the University of Cambridge; 52–53 IWM/Getty Images/Contributor; 55bl, 57t, 61t, 62bl, 62br, 64–65, 65 Walter Nurnberg; 55br, 57b

Photo by Antony Barrington Brown. © Gonville & Caius College, Cambridge; 56–57 War Archive/Alamy Stock Photo; 60 Terence Cuneo/Stills Photography; 63 W Denys Cussins; 67 John Earnshaw; 70b Department of Chemical Engineering; 72, 72–73, 74t The Principal and Fellows, Newnham College, Cambridge; 73 US National Archives and Records Administration; 74t Photos courtesy of D McMullan and K C A Smith; 74br Bell Labs/AT&T; 77tl, 77m, 104b, 109, 112 Haroon Ahmed; 77tr Loop Images Ltd/ Alamy Stock Photo; 79t Keystone Pictures USA/Alamy Stock Photo; 79b Nigel Cubitt; 83t, 83b Cambridge News; 83tr The Shilling Paper; 88 The Computer Lab/Whipple Museum; 91b Keith Thomas; 92–93b Evening Standard/Stringer/Getty; 94 Terry Holloway/The Cambridge Aero Club; 98–99 Phil Sorrell; 105t Julian Eales/Alamy Stock Photo; 108, 186–187 Hugh Hunt; 111tl Tariq Masood; 111bl FCL Photography/ Alamy Stock Photo; 114 Findlay Kember/AFP/Getty; 115bl Keppel Corporation/ Andrew Palmer; 116b Evening Standard/Getty Images; 120 Ronan Daly and Alfonso Castrejon-Pita; 120–121 Harry Coles; 121 Saravanan Balusamy; 122b Judo magazine Vol XI April 1967, No. 7.; 125b Tom Smith; 126 Mohan Munasinghe; 128 Beka Smith; 130tr UIT Cambridge Ltd; 130–131 Crossrail; 131b Ivor Day; 132t Graham Treece; 132b Perry Hastings; 133 Laing O'Rourke; 134, 135b, 136t, 202tr Quang Ha; 135tl, 127t, 152, 153, 154–155t, 156 IfM; 135tr Daniel Strange; 137b Sakthy Selvakumaran; 139t Mike Fallows; 139m Nathan Crilly; 139b James Crosby; 140t Simon Schofield; 142 Catherine Breslin; 143, 166–167 Phil Catton; 144, 162–163 Adrianus Indrat Aria; 145tl Nicky de Battista; 145tr Clare Collins; 146t Rob Gordon; 148–149 Nic Marchant; 154–155b Alex Wong; 158l Mark Welland; 158r Alfred Chuang and Guofang Zhong; 158–159 Rami R M Louca and Yun Thai Li; 159 Tim Armstrong; 160bl Kenichi Nakanishi; 160–161 Ching Theng Koh and Daniel Strange; 161t Jaguar Land Rover; 162t Cambridge Graphene Centre; 162b Kenichi Nakanishi; 164 STAR WARS, (aka STAR WARS: EPISODE IV – A NEW HOPE), from left: Peter Cushing, Alec Guinness, Harrison Ford, Mark Hamill, Darth Vader, Carrie Fisher, C-3PO, R2D2 1977, TM and Copyright ©20th Century Fox Film Corp. All rights reserved/courtesy Everett Collection; 165 CEDAR Audio; 166 Centre for Smart Infrastructure; 168–169 Ian Hosking; 169t Antonios Kanellopoulos; 169b Inclusive Design Centre; 171 Nicoguaro; 172–173 Cambridge Enterprise; 174 Pete Klinger/Alamy Stock Photo; 176, 176–177 Cambustion; 177 Victoria Kronin; 178 Anthony Rubinstein-Baylis; 180–181 Roberto Cipolla; 184–185t Julian Allwood; 189t Cambridge Aerial Photography/Alamy Stock Photo; 189b Grimshaw Global architects; 190–191 AECOM; 192 CSIC; 192–193 Nicholas Hare Architects; 194–195b Ben Seymour and Aya Nawa; 196 Emmet Anderson; 196–197t Mike Kane/Bloomberg via Getty Images; 197 Metail; 199 Ananta Palani; 200l Cambridge University Eco Racing (CUER); 202l Charlie Milligan

CAMBRIDGE UNIVERSITY: NEW ENGINEERING WORKSHOPS.

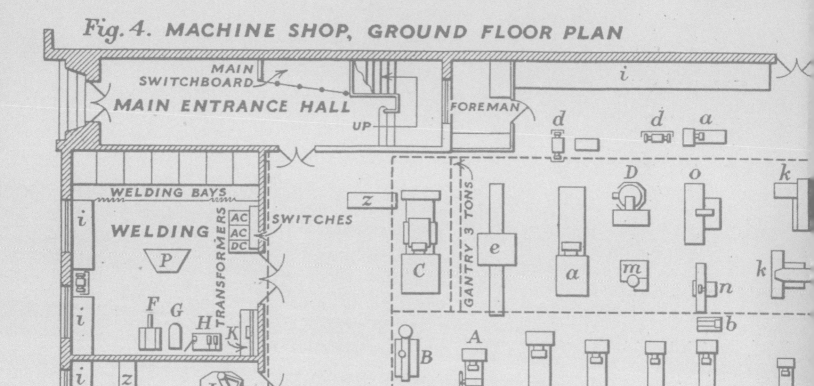

Fig. 4. MACHINE SHOP, GROUND FLOOR PLAN

a	Lathe.	m	Gear shaper.	y	Bending roller.	K	Straight-line cutter.
b	Drilling machine.	n	Surface grinder.	z	Setting-out table.	L	Brazing hearth.
c	Shaper.	o	Plain grinder.	A	Capstan lathe.	M	Engraving machine.
d	Grinder.	p	Guillotine.	B	Radial drilling machine.	N	Buff and grinder.
e	Planer.	q	Nibbler.	C	Horizontal boring machine.	P	Demonstration welding bench.
f	Forge.	r	Mortise machine.	D	Vertical boring machine.	Q	Watchmaker's lathes.
g	Cupboard.	s	Trimmer.	E	Power saw.	R	Coil-winding machine.
h	Linisher.	t	Shears.	F	Spot welder.	S	Slotting machine.
i	Bench.	u	Band saw.	G	Argon-arc welder.	T	Planing machine.
j	Multi-drilling machine.	v	Circular saw.	H	Butt welder.	U	Belt sander.
k	Milling machine.	w	Fly press.	J	Profile cutter.	V	Tool grinder.
l	Universal milling machine.	x	Screwing machine.				